Domestic Building Surveys

D0177376

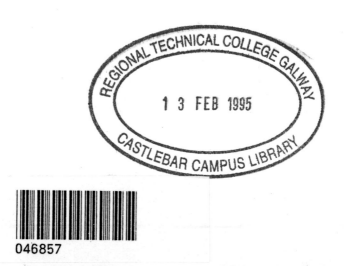

To Geoff Allen and Steve Le Gven for all their help and encouragement, John Hopkins for his able assistance converting my terrible sketches into legible illustrations and Colin . . . gone but not forgotten.

Domestic Building Surveys

Andrew R. Williams

Building Services (Technical)
Chartered Quantity Surveyors
Corporate Building Surveyors
UK

E & FN SPON
An Imprint of Chapman & Hall
London · Glasgow · New York · Tokyo · Melbourne · Madras

692

Published by E & FN Spon, an imprint of Chapman & Hall,
2–6 Boundary Row, London SE1 8HN

Chapman & Hall, 2–6 Boundary Row, London SE1 8HN, UK

Blackie Academic & Professional, Wester Cleddens Road, Bishopbriggs,
Glasgow G64 2NZ, UK

Chapman & Hall Inc., 29 West 35th Street, New York NY10001, USA

Chapman & Hall Japan, Thomson Publishing Japan, Hirakawacho Nemoto
Building, 6F, 1-7-11 Hirakawa-cho, Chiyoda-ku, Tokyo 102, Japan

Chapman & Hall Australia, Thomas Nelson Australia, 102 Dodds Street,
South Melbourne, Victoria 3205, Australia

Chapman & Hall India, R. Seshadri, 32 Second Main Road, CIT East,
Madras 600 035, India

First edition 1993

© 1993 Andrew R. Williams

Typeset in 10/12 pt Plantin by Best-set Typesetter Ltd., Hong Kong
Printed in Great Britain by Page Bros, Norwich, Norfolk

ISBN 0 419 17800 7

A catalogue record for this book is available from the British Library

Library of Congress Cataloging-in-Publication data available

∞ Printed on acid-free text paper, manufactured in accordance with
ANSI/NISO Z39.48-1992 and ANSI/NISO Z39.48-1984 (Permanence of
Paper).

Contents

Disclaimer

The comments made in this book are not intended to be interpreted as necessarily reflecting the opinions of any learned body, institute or institution.

Survey reports are always kept totally private and confidential and they are not issued to third parties. What is more, clients are expressly forbidden to publish or provide copies of reports to others without express permission.

Although all the reports contained in this book are based upon genuine surveys, the location of the office that carried out the surveys has been changed and although still registered, the practice name Building Services (Technical) is no longer in current usage and is not advertised within publications such as *Yellow Pages*.

Names, locations, location plans (where applicable), occupations, trades, and the like, contained within the individual surveys or chapter referring to reports are totally fictitious except where specifically stated. These deliberate changes and 'red herrings' have been introduced to preserve this anonymity and the author and publisher make no apologies for so doing.

In the event of any person choosing to relinquish this anonymity and volunteer information to third parties and thereby creating embarrassment to themselves, their relatives, colleagues or acqaintances, neither the author nor publishers accept responsibility for any loss, consequential loss or expense or discomfort incurred by their so doing.

Whilst every effort has been made to ensure the accuracy of the facts contained within this book, neither the author nor publisher can accept any claims for consequential loss or damage resulting from information contained herein.

Preface

I first contemplated carrying out house surveys when, in my late twenties, friends started buying homes of their own and asked for my opinion. Even though I was a fully qualified surveyor, it was with a certain trepidation that I undertook the first few.

After all, I knew how houses should be constructed, *but* when carrying out a survey, one has to reverse the thought process. It is what *actually* has been built that matters and what defects have been created as a result. At that moment in time, my first points of reference were the files that contained my late father's Building Reports. During his lifetime, he had carried out innumerable surveys for the public. In addition to Building Surveys, his services were retained by a small firm of dry rot and replacement damp proof course specialists.

Like most firms of this nature, they had coloured brochures that illustrated the various types of timber disease found in the UK and the effects of rising damp. I recall that when I had a moment or two to spare, I used to browse through the reports and the files of information, noting the defects and consolidating my knowledge.

In writing this book, I have tried to follow my own experience. The book is based upon six genuine surveys that I have carried out in recent years (copies are in the appendices). I do not claim that these are the last word in phraseology or style. Other surveyors do have different methods of presentation, and I have included a chapter on the recently introduced Report Outlines. The appendices represent the reports in my filing cabinet that I had to study.

Most of the book is deliberately written in the first person singular: to some, it might seem a little pretentious but my idea is that you, the reader, are, as it were, 'accompanying' me when the house in question is being surveyed, as the 'fly on the wall' (or perhaps an assistant)! All you have to do is watch, listen and then question.

The first part of the book hopefully creates a little of the atmosphere, so that you will not be left in a vacuum when reading the later chapters where matters are explained in greater detail.

To save repetition, I have not detailed my every movement whilst in a property. You should assume that procedures described in one chapter have automatically been carried out elsewhere, unless I state otherwise. Each chapter is intended to introduce new problems for consideration and to make you think.

At the end of the relevant chapters in the first part of the book, you are referred to the appendices which contain the reports applicable to those chapters. The contents of each report should then make more sense, in the light of the (more open) comments made before; it is my hope that you will be able to piece together the facts and understand exactly why things have been said in the report. (Although all the cases in this book have been based on actual surveys, the names and addresses have been changed and the locations are totally fictitious. To have done otherwise, of course, would have been unprofessional and unfair to the householders concerned.)

Building Surveys tend to be very personal things. Indeed they bring the surveyor into direct contact with Joe Public. In most books that I have read over the years, this is an element that is largely ignored, except perhaps for a passing comment. It is a pity because, whilst one has to take what the owners of houses say with a certain amount of caution, they can sometimes be a great help when determining possible problems with a property.

In fact, it is worth mentioning that I try to 'involve' the owner of the property before visiting; usually I send a question-and-answer sheet to them, for completion. The sheet covers a wide variety of topics such as the location of the main stopcock; defects and repairs or guarantees of which they are aware; details of extensions/alterations that they or previous owners have made to the property; and the relevant Building Control/Planning references for plans deposited at the local council. There are also examples in later chapters where verbal information, given freely by owners, has been of assistance.

The surveyor carrying out a Building Survey has to cast himself/herself in the role of a detective because that is effectively what he/she is. There are also examples of possible pitfalls (see the Fern Lea report).

I have taken the liberty of presuming that readers may have little know-ledge of domestic building construction and have therefore included sketches and chapters to counterbalance this. I have no doubt that, for some readers, I will have provided too much of this information; but it should be borne in mind that only someone with a sound knowledge of building practices and regulations can effectively carry out a Building Survey. Figure 1 shows the principal parts of a house; and Fig. 2 has been included here, somewhat out of sequence, because it highlights the basic causes of damp in a house. Damp penetration and the subsequent effects on a property are one of the major concerns of a surveyor when carrying out a Building Survey (see Chapter 12 on how buildings fail for amplification).

I make no apology for virtually ignoring non-conventional forms of con-struction, in particular timber frame construction. Although bricks and mortar have their criticisms, historically speaking, system building has not

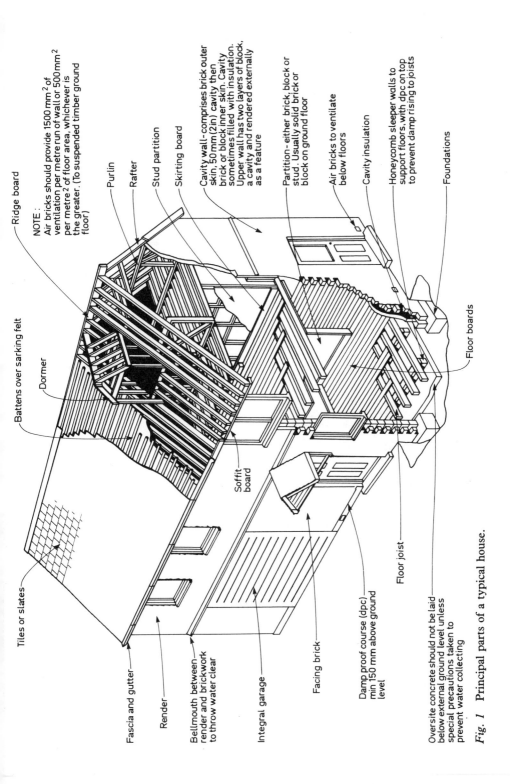

Tiles or slates

Fascia and gutter

Render

Bellmouth between render and brickwork to throw water clear

Integral garage

Facing brick

Damp proof course (dpc) min 150 mm above ground level

Oversite concrete should not be laid below external ground level unless special precautions taken to prevent water collecting

Battens over sarking felt

Dormer

Ridge board

NOTE:
Air bricks should provide 1500 mm^2 of ventilation per metre run of wall or 500 mm^2 per metre^2 of floor area, whichever is the greater. (To suspended timber ground floor)

Purlin

Rafter

Stud partition

Skirting board

Cavity wall – comprises brick outer skin, 50 mm (2in) cavity then brick or block inner skin. Cavity sometimes filled with insulation. Upper wall has two layers of block, a cavity and rendered externally as a feature

Partition – either brick, block or stud. Usually solid brick or block on ground floor

Air bricks to ventilate below floors

Cavity insulation

Honeycomb sleeper walls to support floors, with dpc on top to prevent damp rising to joists

Foundations

Soffit board

Floor joist

Floor boards

Fig. 1 Principal parts of a typical house.

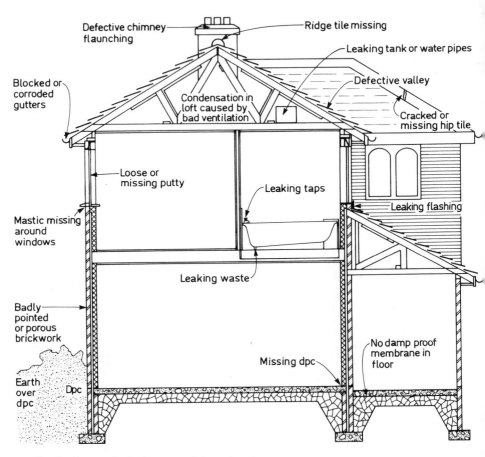

Fig. 2 Some principal causes of damp in a house.

been a wild success in the UK and only time will tell if timber frame structures are suitable for use; I have to confess that I have yet to be totally convinced. The theory of the design might be fine but are the operatives who erect them adequately trained or careful enough?

In some books, the reader is told how difficult Building Surveys are to conduct and the severe consequences to the surveyor if he or she gets things wrong are exhaustively pointed out. To some extent, it is fair to warn of the dangers, but the warnings may be so dire in some cases that if these don't dissuade readers from carrying out Building Surveys for the rest of their days, then the list of dos and don'ts that follows may well do so. Sometimes the check-lists seem endless; I have no doubt that there are many who have hit the 'pain barrier' by page 3! There is too much to absorb too quickly. Here, then, I have tried to introduce information in a logical progression and in an interesting way.

Having said that, carrying out a full Building Survey on a house requires the use of specialized and, in some cases, expensive equipment. The surveyor/consultant who visits the property in question should also be experienced in building construction. For this reason, full surveys should not be treated as another DIY project by prospective owners. (On the other hand, it never ceases to amaze me how many people ask me to carry out surveys on properties when it is fairly obvious that they are in very poor condition.)

Whilst this book is intended for the serious student, it may well be of use to the prospective home-owner who could, with a little knowledge, do his own 'pre-survey' and save wasting his/her money on properties that are obviously highly unlikely to be suitable. When a suitable property *has* been found, the experienced surveyor can be called in.

In writing this book, I also have the hope that I might indeed prove the value of Building Surveys. For most people, having their house surveyed is one of those things that tends to happen because a solicitor, a friend or relation advises them to have one done. If the public knew the sort of things that a surveyor is looking for, and how much money a survey could save them, then they might appreciate both the surveyor's services and the detailed knowledge that goes into providing it.

The final result might also be that they utilize the surveyor's services far more often. Who knows, perhaps one day a system might be created whereby every house in the country has a log book, with surveyors calling every so often to give the house its 'service' and thereby prevent serious defects occurring?

Andrew R. Williams

Acknowledgements

The author wishes to thank Alan Oliver for giving his permission to reproduce his library photographs of dry and wet rot and woodworm (see Plates A to H).

The author would also like to thank British Coal, Pilkington Insulations, Willan Building Services, the Marley Roofing Company and WL Computer Services for their assistance. Many of the sketches contained in this book are adaptations of their literature.

PART ONE: The Surveys

1 We don't call them Structural Surveys any more

Nowadays the term Structural Survey is falling into disuse despite the fact that it is still fairly common for members of the public to ask for one. I know from recent discussions/correspondence with the Chartered Institute of Building, that, in keeping with a decision taken jointly with the Royal Institution of Chartered Surveyors, they are now encouraging their members to use the term Building Survey.

WHAT IS A BUILDING (STRUCTURAL) SURVEY?

It is a detailed inspection of a property by a suitably qualified surveyor, architect or building consultant. The exact extent of the examination is subject to agreement between the surveyor and his client.

Building Surveys can be subdivided into two types, which are:

(a) The destructive survey
(b) The non-destructive survey.

Depending upon the size of the property, and the agreed terms of engagement, the survey can involve just one surveyor or a whole team. A survey of a major building might also require the attendance of building contractors/electricians/heating specialists and, say, lift engineers. (The average domestic survey does not usually involve large groups of specialists.)

I have included six typical surveys in Appendices A to F. With the exception of Appendix A ('The Wetlands'), they were carried out by one person (myself). 'The Wetlands' had a team of four involved.

It should also be obvious from what has already been said that the term 'Building Survey' is rather vague and can mean different things to different people. Therefore, it is essential that the surveyor/practice accurately defines the exact extent of the brief (including all exclusions and inclusions).

Most practices evolve their own terms and conditions over a period of years but recommended conditions of engagement for all types of survey are published by the Royal Institution of Chartered Surveyors (RICS) (Tel. 071 222 7000). It is recommended that you obtain copies and read them.

The destructive survey

The destructive survey is beyond the scope of this book but I will try to give an example of one. Say for instance it is necessary to know the exact construction of a wall so that calculations could be made, then it is possible that instructions might be given for a section of wall to be carefully removed by a competent builder (and rebuilt afterwards) so that the surveyor/ engineer could inspect and take notes.

The non-destructive survey

This type of survey is what this book is all about. This is where a building is inspected and the professional uses his/her knowledge of building construction, building techniques, timber pests and 'on site' observation to establish any likely defects.

The only 'destruction' that he/she will inflict on the property will probably be caused by the use of a moisture meter or drilling holes in cavity walls so that an endoscope can be inserted (these instruments are described later), or by, say, lifting a floorboard.

The major part of the surveyor's skill is interpretation of defects from surface evidence available. For instance, if the surveyor notes that skirting boards (the timbers that run around the edge of a room to stop people kicking the base of the wall) are 'wavy', then his/her natural caution would lead him/her to investigate further in case the distortions were signs that dry rot had set in. (Dry rot is described in detail later.)

The non-destructive survey is most common for obvious reasons.

What would you say to a complete stranger who came to your house and informed you that in order that you sell your property to his client, it was necessary to dig holes in your garden, cut holes in walls and generally disturb the smooth running of your household?

I am sure that your reply would be rude, and ten to one you would refuse to give your permission.

WHAT A BUILDING SURVEY ISN'T

A Building Survey is not a valuation (although a suitably experienced person could value the property whilst undertaking a Building Survey). However, valuing property (i.e. indicating to the client the market value after deduction of obvious defects) is beyond the scope of this book.

In recent years pro-forma reports such as a Housebuyers Report and Valuation (HBRV) have been introduced by the RICS. However, the RICS clearly indicate that this type of report does not carry the same weight as a full (Structural) Building Survey (3rd edition forms still use the term Structural Survey).

Even so, I have included a short chapter on the HBRV.

2　Introducing the headstrong client

Why employ a surveyor to look at a property before you buy it? A lot of people just jump in and hope for the best. Why can't anyone do it? After all, the building society sends someone round. The answer is anyone can, but if I might be permitted, I will relate a short story, the need becomes more obvious.

Some years ago an acquaintance asked me to carry out a survey on a house that his son was thinking of buying. He told me that he wanted the place giving a good going over because he didn't want his rather headstrong son 'buying a pig in a poke'. For simplicity, I shall call the son Mr Green. (Please note in this chapter I am not going to attempt to describe a full survey. I am only trying to prove a point and introduce one or two of the basics.)

At the appointed time, and at his insistence, I duly turned out and met Mr Green at his prospective purchase. I explained what I was going to do and he rolled his sleeves up and volunteered to help me give the place a very thorough inspection. The terraced house in question was empty, by the way, except for some old floor finishes that had been left by the previous owner.

I knew that we had problems as soon as I lifted the old linoleum coverings on the ground floor. The woodworm must have been tunnelling in the floorboards for years and in order to escape to breed they had had to chew their way out through the linoleum but, before they had done so, the burrowing grubs had worked their way along the surface of the boards and the underside of the linoleum, looking for a way out. Consequently, the underside of the linoleum was sitting on pure frass (woodworm droppings) and there were narrow tunnels evident all over the underside of the linoleum. We then lifted some floorboards and found even greater evidence of wood-worm. I could give a catalogue of other defects found but I'm not going to because that is not the point of the story. When Mr Green left, he was singing my praises for finding all the faults and admitted that he was glad that his father had talked him into having a proper survey carried out.

To cut a long story short, I duly prepared my report and set out likely approximate costs for putting the place in order. I heard nothing more.

Four or five years later, another client asked me to survey a terraced house near by.

I seem to have a knack of picking the worst days to do surveys! The sun had been shining all week, but when I went to the house in question, the heavens had opened and rain was lashing down, driven by a strong wind. I found the house with ease and, being slightly early, ran my eyes over the exterior from the comparative comfort of the car, while the rain drummed on the roof.

It was the type of house that had no front garden and, being an end terrace, the pavements surrounded the external walls on two sides. It was also obvious that the house had at one time been two, because a doorway had been blocked up. The area, by the way, looked like something out of the Hovis advertisement, a steep narrow street with each side lined with terraces all the way down. The terraced house in question was at the bottom of a hill. I drove to a position where I could see the gable end.

The reason for doing this is that terraced houses on a hill lean on one another and the one at the bottom of the hill has a tendency to lose its gable end. I examined the wall and noted that the brickwork to the gable *was* newer than lower down the wall. There had been a collapse at some time in the past and it had obviously been rebuilt. How long would it be before the fault recurred? I was fairly confident that it wouldn't be for some years.

When the rain eased slightly, I got out of my car and knocked on the front door. I have the unfortunate habit of rarely forgetting a face but the name nearly always escapes me. When the front door was opened, I recognized the owner and it was obvious that his grey matter was working overtime too. After wracking my brain, I realized that it was Mr Green. Naturally I was surprised, but as the rain started to pelt down again we had very little time to reminisce. Mr Green told me to come in and I virtually dived into the house. It was then that I got my first shock when I nearly fell over. I looked back and realized that the external ground level was higher than the living room floor. It was definitely not a desirable state of affairs but I let the issue of the rather strange floor arrangement slide into the background while I completed the initial formalities. I reminded Mr Green that I had surveyed a house for him some years before. He immediately recalled our former meeting. It came out in the course of our brief conversation that he hadn't bought the first house because of the findings, *but* it was obvious that he hadn't learned very much.

Despite what he had said at the time, instead of appreciating my services as he had professed to do, it seemed rather that Mr Green had been annoyed that his father had interfered with his intended purchase. In a fit of pique (he didn't use that exact word) he had gone off and bought his present home without consulting anyone.

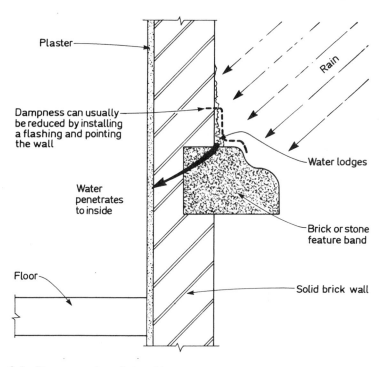

Fig. 2.1   Damp entering via band in wall.

With introductions and explanations completed, I deliberately deviated from my normal sequence of carrying out a survey, because of the rain. (Normally, I like to view the house in question from the outside first so that obvious defects like missing slates can be located when I inspect inside.) I justified my actions with the thought that, if I went into the roof void during one of the heaviest downpours there had been for quite some time, any major defect was sure to become obvious. After going through the roof space, which was in reasonable condition because it had been re-roofed (and then cleaning myself up before walking through the rest of the house), I then started examining the bedrooms. Dampness was immediately apparent on the external walls. In all rooms, the wallpaper was very badly stained. After a while, and after staring out of the upstairs windows to locate a likely cause of the staining, I came to the conclusion that a projecting band feature course externally was directing water into the house and staining the plaster work internally (Fig. 2.1). This discovery, however, was only the start. I took a moisture reading (see later for description of moisture meters) on the gable and was surprised how low it was. I had anticipated levels being recorded similar to the rest of the main bedroom. Upon tapping the wall, I found it rang hollow. Then I noticed that the skirting board (see Fig. 1) was a completely different timber section when compared with the rest of the room.

I asked Mr Green's permission, rolled back the carpet and then lifted a set of short floorboards that had quite obviously been made into a sort of mini trapdoor. (According to Mr Green, an electrician friend of his had cut the boards when he had installed a spur on the mains circuit.)

Once this was done, my worst fears were substantiated. As I suspected, someone, maybe Mr Green himself, had lined the gable end to create a cavity and prevent the decorations becoming stained. This type of lining, however, whilst disguising the fault, creates the ideal conditions for a dry or wet rot outbreak. As suspected, moisture meter tests taken on the joists and timbers hidden from view gave very high readings. (*The moisture levels were building up in the hidden void and creating a good breeding ground for infestation to set in . . . See later chapters.*)

During the course of the survey, I uncovered the following defects.

(1) As described above, the gable wall on both floors had been lined with battens and plasterboard in an attempt to disguise the penetrating damp that was affecting the wall.

(2) As I said before, the house had obviously been two at one time and had been knocked into one. When I lifted the carpet covering the junction between the two original properties the floor was very irregular and the adjacent boards were very damp. The top of the former party wall was also visibly damp. I came to the conclusion that the damp course was either non-existent, defective and/or bridged and damp was rising. I told Mr Green, in very mild tones, to avoid giving offence, that it would be sensible to let me lift a floorboard to see what we found. It all had a terrible sense of *déjà vu* about it.

He decided that he wasn't very happy with what I had found either. I wasn't sure whether I believed him or not but Mr Green claimed that he had never looked under the carpet since he had moved in.

I deliberately let the matter drop. Whilst it may give the surveyor joy to think that he has earned his corn, I could tell that Mr Green was far from happy. I could almost hear his brain ticking over as we worked. He knew that the repairs were going to cost dear. To give him his due, he set about lifting boards with a vengeance. By the time he'd finished, we had about ten up and it was possible to view almost half of the subfloor area in the left-hand front room.

As the boards came up one by one, the truth came out. Whoever had formed the archway between the living rooms of the two original properties had pushed all the demolition material into the floor void instead of paying for skips to take the debris away.

As far as I could tell, judging by the number of air bricks in the walls outside (Fig. 1), the subfloor always had been badly ventilated but blocking up the flow of air with rubbish is bad practice and undoubtedly harbours diseases. Within minutes of exposing the timbers we found signs of rot where the joists were built into the outer walls. Then I found water streaming into the subfloor void from somewhere. The reason is answered below.

(3) I found that an air brick in one part of the house was at pavement level. Water running down the street was actually flowing through the perforations and under the floor of the house, further dampening the timbers. After continuing the inspection for some time, I came to the only conclusion that I could. It appeared that sometime in the recent past, council workmen had very helpfully relaid the pavement. Instead of building it at its original level they had lifted it in that area of the road by approximately 1 ft (300 mm). It certainly provided answers as to why you had to step down into the property.

Naturally, in damp weather, water tended to seek the lowest points, which were Mr Green's living room floor and the subfloor under. (Floors, by the way, should be 150 mm (6 in) above ground level and not 300 mm (1 ft) below it).

(4) The external walls of the house badly needed repointing. When I probed the mortar, it was soft and had been worn away over the years. The need to repoint was proved internally by the numerous damp patches on the wallpaper.

NB Solid walls will let in penetrating water far more easily than cavity walls and bad pointing only increases the risk of damp penetration.

(5) The roof had been re-tiled using heavy concrete interlocking tiles. The original coverings, judging by the surrounding houses, had been thin slate. Upon inspecting the roof internally, I found signs that the rafters probably were overloaded. The loft space also had very little ventilation because what looked like newly laid glass fibre insulation was blocking the air gaps at the eaves. I came to the conclusion that ventilation tiles (described in later chapters) needed installing. These tiles are designed to let moisture out without letting rainwater in and are now becoming very common in usage.

(NB The new Building Regulations require that most roofs should be thoroughly cross-ventilated. The amount of cross-ventilation required depends upon the roof type and pitch. Glassfibre and rockwool quilts have to be kept back so that they do not obstruct the ventilation gaps at the edge of roofs. Patent under tile vents are now available to solve this problem. This topic is discussed in greater detail later on in the book.)

As a consequence of lack of ventilation, condensation was forming in the roof void of Mr Green's house. Moisture meter readings on all timbers were found to be well in excess of the recommended limits. In time, unless roof ventilators were put in, rot would undoubtedly attack the roof spars, weakening them still further.

So, to the moral of the story: Mr Green had saved himself the cost of a Building Survey when he bought the house but the cost of rectifying the defects would probably have cost him some *50 times* that amount in repair bills (unless he found another person who would buy the property without a survey). It was definitely a case of penny wise and pound foolish.

3 Dampness at Shepherd's Cottages

My next short story is meant to illustrate that the surveyor must not jump to conclusions. It also amplifies some of the comments made in the preceding chapter concerning condensation. For the sake of a name, I shall call my next client Mr Cornwallis and the village in which he lives Shepherd's Cottages.

It seems like a nice friendly name for a small village, doesn't it? However, I have named the area with my tongue rather firmly in my cheek. In my experience, if a place has a nice name, then there is just a chance that some developer is trying to pull the wool over someone's eyes!

Now I'm being cynical. Let's just say that despite the delightful name of the place, as far as I am aware there hadn't been a shepherd, let alone sheep, anywhere near the place for over 150 years. But that is by the by.

When Mr Cornwallis rang me, two years after I had done some work for him, I was somewhat concerned. Perhaps I'd better explain.

His house was a relatively modern town house. The house was fairly conventional but the front and back walls of the house were built in what has become known as 'timber frame construction'. (Without going into the technicalities, Fig. 3.1 shows the basic type of construction.)

The walls between the town houses were brick-built cavity type and were the main structural walls for all the houses in the terrace. From what Mr Cornwallis told me, I gathered that the houses were not a resounding success because there was a general problem of condensation in the whole block. Most of the people who had lived in these houses for some years had replaced the timber-frame parts of these houses. Apparently they had found that the timbers rotted very quickly. Those that didn't rot became covered with black and green mould. Whilst this was not deleterious to the timbers, it looked unsightly.

NB To be fair to the modern timber frame manufacturers who nowadays build houses to very high standards, the structures at Shepherd's Cottages were a poor imitation, lacking vapour barriers and adequate insulation. All that the builders had done was install massive window panels across the front and rear of the

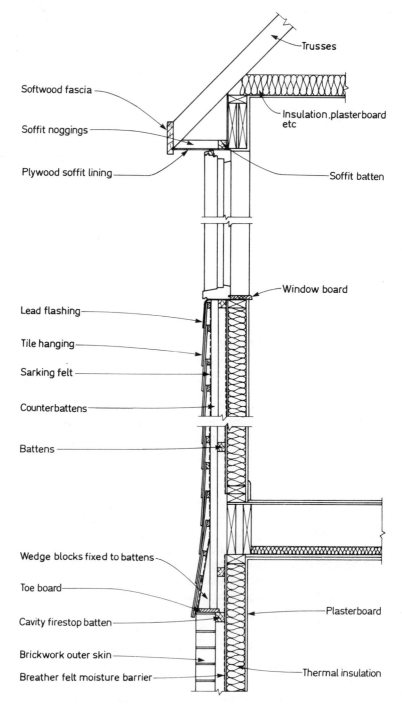

Softwood fascia

Soffit noggings

Plywood soffit lining

Trusses

Insulation,plasterboard etc

Soffit batten

Window board

Lead flashing

Tile hanging

Sarking felt

Counterbattens

Battens

Wedge blocks fixed to battens

Toe board

Cavity firestop batten

Brickwork outer skin

Breather felt moisture barrier

Plasterboard

Thermal insulation

Fig. 3.1 Typical timber frame section.

properties. (These houses were built prior to the introduction of energy saving clauses in the Building Regulations.)

As I said, most people had replaced these walls and Mr Cornwallis wanted to follow suit. His brief to me, two years before, was to prepare plans which showed the 'timber frame' parts removed and replaced with conventional brick and thermal block outer cavity walls and the windows replaced by double glazed uPVC type. (I will describe general building construction later on in the book.)

I duly submitted plans to the council on his behalf and received approval.

Now, as I said, when Mr Cornwallis rang me, I was somewhat concerned. Two years had elapsed and now he was telling me that the damp problem had returned worse than ever. Whilst I knew that my design was to modern standards, implicit in his phone call was a note of criticism. I had failed him!

I duly visited Mr Cornwallis's house the following day. He invited me in and then told me that he thought the house had rising damp and that this had been the problem all along. He then went on to tell me that if I had told him this in the first place then he wouldn't have wasted his money rebuilding the front and back walls. Even though I knew that I had done nothing wrong, it was an awkward situation.

I asked him how he had decided that he had rising damp. He told me that he had called in a firm of damp proofing contractors to diagnose his problems.

I asked for their name and he told me. From experience, I knew that the firm concerned were far from reputable. Mr Cornwallis continued. They told him that an injection damp proof course in the gable would solve all the problems. However, before doling out any more money, he decided to consult me first.

He asked me to test the walls with my damp meter and see if I got the same readings as the damp proofing company. I did so and the meter duly shot off the dial. Cornwallis seemed relieved. At last, he had his solution! All he had to do was to get the damp proofing firm to inject his wall and the world would be his oyster!

However, despite the fact that Mr Cornwallis had his solution, I wasn't happy and went around making more tests. I took a series of readings with the meter (I describe a moisture meter later on in the book) in the lounge. Every reading was high. I then went into the kitchen, weaving my way through containers of nappies, and took further readings.

I started to suspect the obvious – condensation again – I decided to go upstairs. They say that you can't test for condensation using a standard Protimeter. It is, in theory, not designed to do the job. However, when condensation is very bad and when the walls are lined with vinyl paper or painted with gloss or eggshell paint, the thin layer of water can be detected merely by gently placing the 'vampire' prongs of the meter against the surface. Upstairs, I duly did so and got a high reading. Being a surface reading, the damp could not be from the plaster behind. Added to this, it is almost impossible for rising damp to be present some ten to eleven feet (3–3.5 m) above ground level. I concluded that the problem was condensation. Upon looking behind a wardrobe, I found mould about an inch thick which confirmed what I had thought. Mr Cornwallis had his condensation problem back with a vengeance! However, I knew that he was unlikely to believe what I told him without some sort of very good explanation or proof. He was the type of chap who perhaps would throw any amount of money at a problem without really considering his options and it was obvious to me that the damp company had his ear at the moment. All I had to do was to tell him what he wanted to hear and he would rush to the telephone and call in the injection team.

I was equally convinced that rising damp was not the problem. Where was all the moisture coming from? Then I thought about the nappies again. I went back downstairs and asked Cornwallis's young wife if she had a tumble dryer. She smiled and said 'yes'. Her doting husband had bought it for her about four weeks before. I then asked if she had a flexible hose to put out of the window. She shook her head. It turned out that the tumble dryer was situated in the middle of the house and all that she did was open the back door when using it. If you have seen the amount of steam that these machines push out whilst working you can guess that this was the cause of the problem. To make matters worse, the house had an open staircase very close to where the tumble dryer was situated. A large amount of the steam undoubtedly went straight upstairs. I ran my eyes around the kitchen. There were nappies everywhere. Judging by the number, the machine must have been run several times a day.

Mr Cornwallis seemed unconvinced when I told him the reason for his problem. It was all too easy! He agreed, however, to resite the tumble dryer and vent it externally. Two weeks later he rang me. After some thought, he had taken my advice and now that a few days had elapsed, the damp had cleared up.

It is worth noting this rather extreme story because condensation is becoming a major problem in modern housing. The 1985 Building Regulations (and 1990/92 amendments) have recognized this and ventilation has become a major consideration. Don't overlook condensation as a possible cause of dampness.

Note
In 1993 Protimeter PLC (Tel 06284 72722) introduced their 'Interpro' digital condensator which is designed to measure/evaluate condensation problems in a building.

4 'The Wetlands'

(For full report see Appendix A: 'The Wetlands' report)

Note
(See Fig. 4.1 for sketch of property and surroundings . . . This sketch was made prior to carrying out the survey. When dealing with large properties a sketch of the building can save a great deal of writing.)

In this chapter, I intend to describe a full survey. The one at 'The Wetlands', in Tipham (the names have been changed once again), was an interesting one because it involved myself, an engineer, a plumber and an electrician. For the want of a name, I shall call the client Carl Vernon.

When Mr Vernon telephoned me, I have to confess that I didn't know the man from Adam, but then with domestic building surveys it tends to be like that. I suppose that it is not really all that surprising. After all, the average person changes his/her house only a few times in a lifetime. When they do, quite often it will be caused by a change of job and will result in a move from the area in question. So it is unlikely that the same person will return very often, if at all.

Vernon asked me if I could report on a property that he was thinking of buying in Church View. He told me that I had been recommended by a friend, but to this day I don't know who did the recommending.

I discussed his requirements in as much detail as it is possible to do over the telephone and tried to ensure that he fully understood what I did, and did not do, during a survey.

It is essential that this is done because, to the general public, the term 'Building Survey' (or as most people refer to it, 'Structural Survey') can mean a hundred different things. From experience, I find that it is essential to establish the amount of work to be carried out and to try to set a budget fee. Some people live in cloud-cuckoo-land and it is essential that they know what their liabilities are going to be so that the possibility of a dispute arising over fees is obviated. However, it is essential to know the exact extent of work prior to the fee being agreed because the amount of work in a full survey can vary so much.

Eventually, we settled on a likely charge (after he had described the property in detail, and what he wanted me to do). He seemed partially

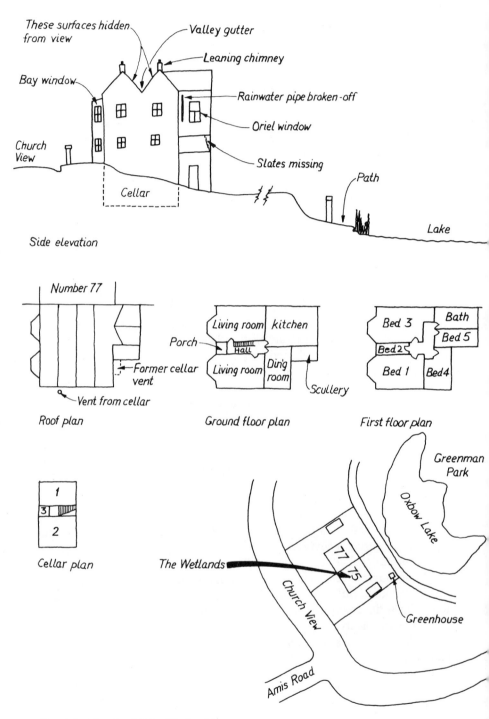

Fig. 4.1 Sketch of 'The Wetlands'.

satisfied but then noted that I made only visual inspections of the electrics, heating system and drainage. He particularly wanted to have the electrical system and the heating systems checked out.

I told him that I could arrange for an electrician and plumber to accompany me if he wished but naturally I would have to add their charges to my account. I also advised him to allow me to obtain a Mining Report. Tipham was, after all, a mining area. I also explained that as British Coal charge for this service that I would also need to add this cost to my account. (Figure 9.1 shows a typical Mining Report.)

It was then that Mr Vernon hedged. He told me that he would call me back after he had chatted the matter over with his wife. Vernon did not telephone; instead, he called in at the office. As luck would have it, I was in.

Vernon had brought an estate agent's leaflet with him and asked if I could act for him. It was obvious that he had telephoned several other surveyors in the area, prior to his visit. Presumably, he had come to realize that my quote was not excessive.

He told me that he had had a preliminary look around the house himself and it seemed ideal. I gained the impression during our conversation that Mr Vernon had lived and worked in London for most of his life but since retirement wanted to return to his roots.

He gave me the estate agent's sheet and I made a note of the address. I also asked for his present address and phone number. (Such details are obviously essential, even if only to know where to send the bill.) I knew the area in question because I had been called in by someone else in that location in the recent past. From what Vernon said, and judging by the estate agent's sheet, it was quite a sizeable property.

Vernon then told me that one of the things which concerned him was evidence of some subsidence. I told him that I would obviously check the property out for such things as woodworm, dry rot and the like but in *bad* cases of subsidence a full engineer's report would be needed.

Vernon didn't think that this would be necessary, but we agreed that I could take my engineer with me. His duties would be restricted to the matter of possible settlement and the inspection of a tie in one of the external walls and I would check the house out for other defects. (Typical ties are illustrated in Fig. 4.2.) It was a common practice at one time to stabilize bulges in walls by running a large diameter metal rod through a building and then placing a tie plate externally. The tie then prevented further movement.

After that, we went through the normal routine. I took down the details of the present owner, because naturally one cannot just enter a property without permission.

It turned out that the house was unoccupied and that the keys were in the hands of the estate agent. However, the current owner's relatives wished to be present when I carried out the survey. I took down their name and address and said that I would notify them of my intentions.

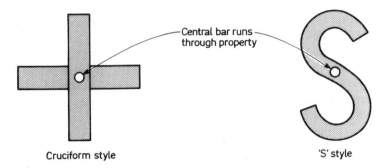

Fig. 4.2 Typical ties.

However, before doing so my first task was to round up my 'pet' electrician and plumber. When I visited the property, I wanted everyone there so that we pulled together as a team. Luckily for me, both of them made themselves available and said that they could visit the house with me four days later. With that sorted out, I started to follow my normal work pattern and contacted the estate agent. I made the arrangements to pick up the keys and immediately drafted out a standard letter to both my client and the estate agent cofirming the arrangements. I also sent a letter to the relatives and enclosed my information sheet.

At the appointed day and time, we visited the house. (*Note I had visited the house with my engineer two days earlier because he was unable to visit the house with the others. Although there were slight signs of settlement, we were both satisfied that the hairline cracks were of no real significance and that the tie seemed adequate. Our only real concern was a leaning chimney stack.*)

The house was a rambling semi-detached property not far from a local lake and, no doubt, the sight of ducks and swans floating around on the water wagging their tails in the air as they dived for weed barely two hundred yards from his proposed new home appealed to Vernon. However, I knew from our discussions that the house had a cellar, and as one can imagine, lakes and cellars should bring doubts to anyone's mind. I was particularly concerned because the person that I'd visited not far away, only a few weeks before, had been complaining that his cellar seemed to flood at various times.

Note on cellars
As a matter of interest, when I made enquiries with the local Water Board, I was told that the water tables (level of underground water) in the whole of the Tipham area were rising because as local industries closed down (caused by the recession), the amount of water being pumped out of the ground was decreasing. For the first time in years, water levels were rising. From what I was told, I had no doubt that cellar dampness would become a greater problem in future years.

I tend to use a voice activated tape recorder (Tandy produce a good machine) to take survey notes because I find that it helps to save a great deal of writing whilst actually on site. Being voice activated, you use up tape only whilst actually speaking. As soon as we arrived, I made a note of the job, the time of commencement and the weather conditions. Weather conditions can be important at a later date because it's obviously far easier to detect certain problems in bad weather than, say, during a heatwave and if queries do arise a note of the conditions is a useful record.

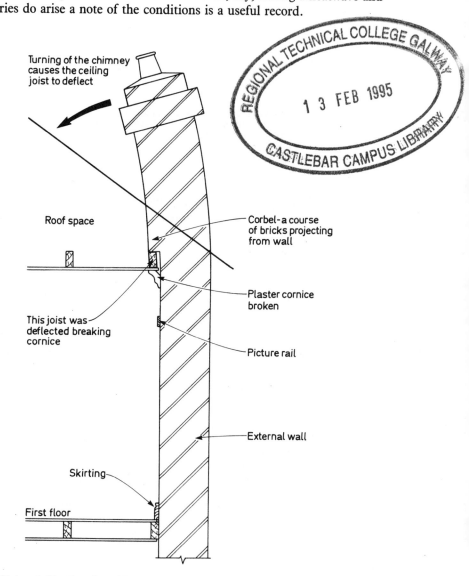

Fig. 4.3(a) and (b) Leaning chimney.

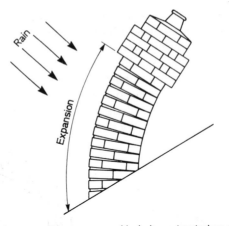

(b)   Leans on chimneys caused by being saturated on one side
for a long time - sulphates from flue gasses attack mortar
causing bending away from the normal wind/rain direction

Fig. 4.3   Continued

Then, whilst the electrician and plumber went to find the meters and stop cocks, I made an outline sketch of the building (Fig. 4.1).

I did this because, being a fairly large property, I needed to gain my bearings. After completing the sketch, I then went outside to form an overall impression of the property. (My previous visit had lasted only a few minutes.)

During that time, I noted four or five obvious faults which I dictated on the machine.

For the record, what I noted was as follows:

(a) One chimney was leaning inwards towards the house (*this item was discussed previously with my engineer*) and, if left, I had no doubt that the new owner would find it crashing through into his bedroom at some future date.

(b) There were slates missing off a verge. (A verge is the edge of a slated or tiled roof that overhangs the gable.)

(c) Ridge tiles were missing in two locations.

(d) Moss was growing on walls where rainwater pipes were broken off.

(e) There was a crack in a cove in an upstairs bedroom.

Admittedly the notes were not in a logical order, but at least I had recorded the problems and they would not be forgotten.

For the time being, merely noting the defects was sufficient. The two rainwater pipes that terminated half-way up a wall had obviously been missing for some months because moss was growing in streaks down the wall surfaces. The disrupted coving in one of the bedrooms is worth a special mention. I had yet to connect the leaning chimney stack outside and the damage internally and it was necessary to spend quite some time in the loft space later on in the survey to satisfy myself regarding the cause (Fig. 4.3(a)).

It was at that point that an overflow pipe started to discharge. (An overflow pipe is a short length of pipe that projects from the property which provides a visible discharge of water if a tank or the cistern on a toilet is in danger of overflowing.) I guessed that the plumber working inside had set it off when he'd turned on the water.

I went inside and told the plumber to turn the main stop cock off again until he'd managed to locate the tank and find out why the overflow was cascading. From previous experience, I immediately suspected that the ball valve in the loft was defective and that the previous owners had turned the water off prior to leaving in order to prevent the defect from continuing in their absence.

I then started the survey in earnest. I like to follow a pattern when carrying out a survey. Consistency helps to avoid missing defects. I usually

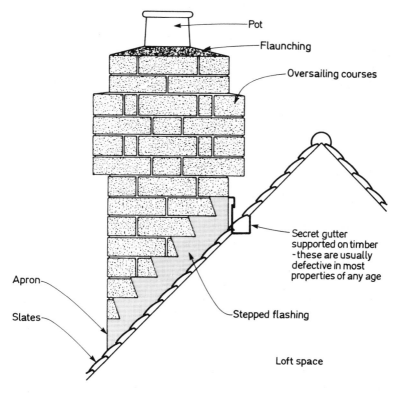

Fig. 4.4 Chimney details.

start by examining the roof coverings, flashings, gutters and chimneys with the small binoculars that I carry. (*NB* High magnification binoculars are to be avoided as it is difficult to focus properly over short distances – the opera glasses type are better.)

The main roof was in two distinct parts with a valley between the two (Fig. 4.1). To the rear of the property was a lower roof over the rear 'outrigger'.

I had been told by Vernon that the high level pitched roof had been 'overlaid' with bitumen and hessian.

The mere fact that the roof had been overlaid in this way made me suspicious. The roof was probably becoming what is known in the trade as 'tired'. That is the slate nails had started to rust and slippage had started. By bonding all the slates into one mass they are prevented from slipping, but undoubtedly before this work had been carried out the roof must have had leaks. (Owners of property have a nasty habit of ignoring preventive maintenance and it is only when something out of the usual happens that they take action.)

Where the roof was not covered in felt, slates were missing and flashings were either loose or missing. (Figure 4.4 shows chimney and flashing details.) All the chimneys needed repointing.

I broke off at this point because Chas, the plumber, appeared and informed me that he'd found something nasty lurking near the water tank. I made a note on the tape as a reminder and then went inside and was presented with a section of the timber cover that had been over the tank in the loft.

Apparently the ball valve in the tank (Fig. 4.5) had been defective for years and the previous owners had not bothered to have it repaired. They had merely wired it up in an attempt to stop the tank overflowing. The water level had inevitably risen during periods of the day when the family were not at home and, in the humid and unventilated space over the tank, rot had set in. I thanked Chas and told him that I would pay special attention to that area when I went inside.

After completing the external roof inspection, my next task was to check out the gutters, downpipes, walls, windows and fascias on each elevation noting carefully what I found. This I carried out working off a short set of ladders that I had loaded into the car. Obviously it is not possible to check out high level areas with a short set of ladders, but most surveyors don't carry long ladders and state this in their conditions of engagement. Whilst this may seem to be dodging the issue, one normally finds that when timberwork is defective at ground level, that the condition of higher level timbers is unlikely to be any better. Besides, it is possible to examine upstairs windows and some timbers by opening the various casements or sashes and reaching out. I found that the windows and timberwork

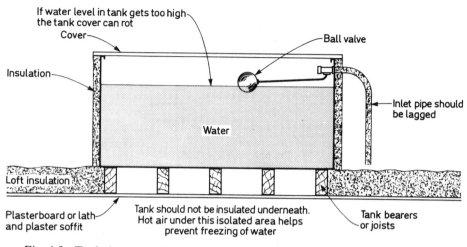

If water level in tank gets too high the tank cover can rot

Cover

Ball valve

Insulation

Inlet pipe should be lagged

Water

Loft insulation

Plasterboard or lath and plaster soffit

Tank should not be insulated underneath. Hot air under this isolated area helps prevent freezing of water

Tank bearers or joists

Fig. 4.5 Typical water tank.

generally were in poor condition. I carry a sharpened screwdriver for a probe and a moisture meter (a moisture meter used wisely can be a surveyor's best friend) and jab the timbers at regular intervals. In this particular case, my probe slid into the woodwork at least 2 in (50 mm) on most of the windows, which obviously wouldn't happen with sound wood. The areas that I paid special attention to were those where water lodges the most (cills, junctions between the cills and the uprights and the glazing bars). That exercise completed, I examined the walls generally. I quickly discovered that the house didn't appear to have a DPC (damp proof course) or, if it had, someone had buried it by raising the path levels around the house. (For those who don't know, modern houses have DPCs incorporated around the whole house at approximately 6 in (150 mm) above ground level.) I always make a special point of checking the DPC because defective, non-existent or 'bridged' DPCs are one of the most common faults in any property and burying DPCs by building flower beds or raising path levels is a 'crime' that a large number of home owners commit without realizing that they are creating problems for themselves or future purchasers.

From ground level inspection the gutters at the rear of the property appeared to have been replaced by a new PVC system but the downpipes had never been renewed. I carry a rather nifty device which is basically a mirror on a stalk and it is marvellous for looking behind pipes.

The downpipes were the old cast iron type and set away from the wall with 'gas barrel distance pieces' (Fig. 4.6).

Upon inspection I discovered that the downpipes had, as I suspected, never been painted at the backs. It is something that most people overlook

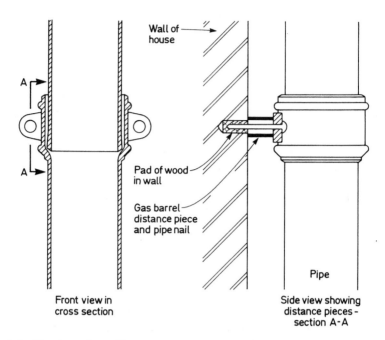

Wall of
house

A

A

Pad of wood
in wall

Gas barrel
distance piece
and pipe nail

Pipe

Front view in
cross section

Side view showing
distance pieces -
section A-A

Fig. 4.6 Cast iron pipe with ears.

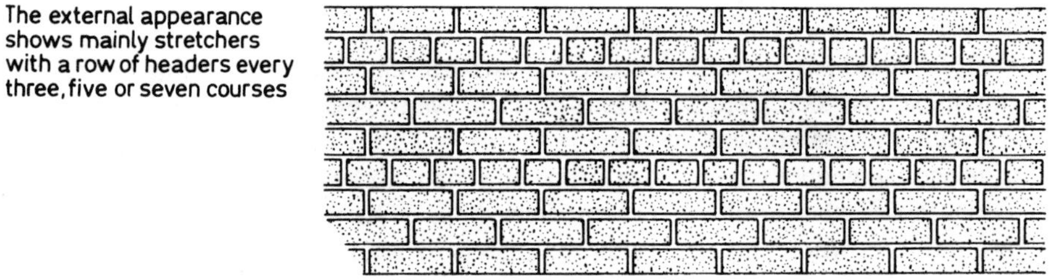

The external appearance
shows mainly stretchers
with a row of headers every
three, five or seven courses

Fig. 4.7 English garden wall bond.

when they are painting a house. It's a case of 'out of sight, out of mind'. As a consequence, most of the downpipes were rusted through. In this particular case, this defect was made all the more noticeable by the moss that was growing down the walls adjacent to each pipe. I also found that one downpipe was split and that a gash of some 10 mm wide and 100 mm long was apparent in the mirror. In wet weather, it no doubt discharged rainwater directly onto the brickwork. I made a note to inspect the area inside very carefully because it was an area where one could anticipate real trouble caused by water penetration, especially as the walls appeared of solid construction; they were contructed in English garden wall bond (Fig.

4.7; other brick bonds are detailed in later chapters). In the main, the use of English, Flemish or various types of garden wall bond indicates that the walls are of solid construction and therefore most likely to suffer from damp penetration.

NB Solid walls are precisely that, they are solid as opposed to the newer cavity wall which is built in two independent skins, tied together with wall ties.

It was about this time when Vernon arrived; I suppose it was natural that he wanted to see what was going on and make sure that his surveyor was earning his fee. After he had found me, he asked me what I thought of it so far. I resisted the obvious answer, in case he took the remark seriously. I just told him that there was no DPC.

Some people object to their client being present while a survey is being carried out but I make a point of letting them attend if they so wish. At least, if they can see what you are doing, they can appreciate how time consuming Building Reports can be.

The other point in its favour is that whilst you might be doing your best to detect all the faults, quite often they have visited the property on several occasions and do sometimes ask questions about 'faults' which can be beneficial inasmuch as you can put their minds at rest about certain features – on the other hand the question can act as a trigger. You can then make a special point of checking out any item that they are concerned about and discuss worries, whether real or imagined. The wise surveyor checks these out.

He asked me if the damp proof course was important. I explained that without one the house would suffer from rising damp. I also explained that as the house had solid walls the problem would be accentuated. He asked me how I knew the walls were solid and what should they be.

I explained that the walls were built of English garden wall bond (three stretcher courses to one header course) and (unless the builders for some reason known only to themselves were using 'snap headers') that the walls were almost certainly solid. I also explained that I had measured the thickness of the wall and that it didn't seem thick enough to be a cavity wall. It was soon obvious that Vernon had absolutely no knowledge of building construction at all, so I explained that the modern form of external wall construction was a cavity wall and that this, in the main, was better than a solid wall because it kept water out better.

I then used my moisture meter to indicate that some damp was travelling up the wall.

After a while Vernon became bored with what I was doing and drifted off to interrogate the electrician and plumber.

Shortly afterwards, I was at the side of the house investigating a rusting old oil tank when I found a 6 in (150 mm) pipe sticking out of the ground. It seemed to serve no purpose but then, on reflection, I decided that it might be ventilating the cellar. However, it was also obvious that the original pipe had had some sort of capping and this had been missing for some time. Obviously, an open pipe at ground level would act as a funnel, and send water into the property. I also checked the render plinth that ran around at ground level. At this point Vernon returned and started asking questions about the plinth. I explained that this was a rather outdated method of trying to prevent rising damp.

Once I had completed outside (except for surveying the grounds), I changed into a stout paper 'jump suit' that I always carry with me and prepared to enter the loft.

In order to check out areas like lofts, I always carry a powerful 'lead lamp' and a roll of cable to connect to a power socket. I find that torches are all very well but to inspect lofts properly a great deal of wattage is required.

The trapdoor into the rear loft was in the bathroom. The bathroom, in turn, was housed in the 'outrigger' section of the house (the part of the roof that didn't have bitumen and hessian on it). The previous owners had created a secondary ceiling in this area and had provided an access trap so small that it was barely possible to climb through. Above the false ceiling I found the original old trapdoor and opened it. My first job was to look at the cover that the plumber had found. My inspection proved what I already suspected. The wood had been in such close proximity with the water in the tank that wet rot had eaten most of it away. I duly recorded the fact and then searched the area for any other signs of wet or dry rot. Luckily there were none. I did, however, in my searching find an outbreak of active woodworm. The fact that it was active was proven by the piles of 'frass' (woodworm droppings) apparent all around the area.

I was glad of my overalls. The loft space was filthy and had soot deposits covering every flat surface. As you will probably gather from what I said before, access to the loft was difficult but getting into the main roof areas was a pig of a job. The original bathroom ceiling was some three feet below the main house ceiling and in order to enter the main loft you had to literally climb over the water tank that some 'helpful' plumber had seen fit to fix in the most inconvenient position possible. Before entering the loft, I had noted that the hall below had a lay light in it (a horizontal pane of glass flush with the plastered ceiling). This lay light then had a vertical framework that ran up until it hit the roof and above that there was a roof window. I struggled over the tank and wormed my way past the lay light framework. Mr Vernon briefly poked his head into the space above the bathroom but decided that he would stay where he was and contented himself with inspecting the rotten timber that had been removed from the overfull water tank.

I would at this point make special comment about lofts. I know that most people will accuse me of 'trying to teach granny how to suck eggs', but I am going to state the obvious. Lofts can be dangerous places. Unless the loft is fully boarded, don't step on anything other than a structural timber (e.g. ceiling joists). Most ceilings are either made of lath and plaster or plasterboard which is nailed to the ceiling joists. Ceilings will not take the weight of a grown man or woman!

To get back to the survey, I found a suitable safe place to stand and I turned off the lead lamp for a few moments. After letting my eyes adjust to the darkness, I ran my eyes over the inside of the roof. I was pleased to note that I couldn't see any signs of daylight, except around one chimney stack.

It was obvious that the hessian and bitumen layer was intact and that the only place where light could be seen was where a lead flashing had come adrift.

I then turned the lamp back on and went through my usual procedure. Firstly, I inspected the floor of the loft. It was insulated with approximately 25 mm (1 in) of glassfibre. Modern regulation requires that quilts should be far thicker. Under the new amendments to the Building Regulations, a quilt of 150 mm (6 in) thickness is now the norm for new houses. I made a note of my findings and then inspected the ceiling joists in more detail. Naturally I was particularly concerned regarding woodworm spread. I tested the timbers in several locations and found the moisture levels very high. My suspicion at that point in time was that the hessian and bitumen were preventing air flow in the loft and consequently trapping moisture. (As you will note from my report (Appendix A), I came to the conclusion that ventilating slates needed installing, Fig. 4.8.) Condensation in loft spaces is becoming a common problem, because as insulation thicknesses increase, without more ventilation being provided, the risk of condensation also increases.

Once I had finished my general inspection of the timbers, I then went to find the chimney that had a lean. It was then that I discovered the reason for the damaged coving in one of the bedrooms. The chimney breast inside

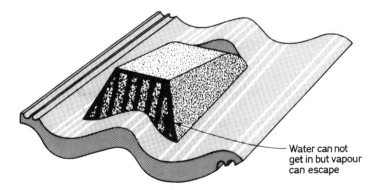

Water can not get in but vapour can escape

Fig. 4.8 Typical roof ventilation tile (various types available to suit slate or tile profiles).

the loft was wider than the external wall of the house and the wall had been corbelled out (Fig. 4.3). The lean had forced the corbel into a ceiling joist which had bent and broken the coving below.

My next real concern was the valley (see Fig. 4.1 for location). While secret gutters and valleys are acceptable building construction, they have an inbuilt weakness. If the lead perishes then the surrounding timbers will start to be dampened by rainwater. Once damp, fungi can take hold. Despite the dust, I tried to crawl as close as possible to the base of the valley and secret gutter and was rewarded by finding a dry rot outbreak in one location. Perhaps I had better amplify at this point. Dry rot is, in my opinion, one of the worst defects that houses in the UK can suffer from unless checked in the early stages. The fungus is to houses what cancer is to humans. Given the right conditions, dry rot can spread extremely rapidly and will drive its tendrils behind plaster and through brickwork. It needs a specialist to kill it. Dry rot has several forms during its lifetime but the most obvious is the 'fruiting body'. (Dry rot is described in later chapters.) This is the reproductive organ and sends out spores as fine as dust all over the place to infect other areas. (If you require further details, leading treatment companies produce excellent detail sheets which they are usually only too willing to supply.)

I made a note on my recorder that the client should contact a firm of dry rot and woodworm specialists in order to obtain a full survey of the extent of the various attacks and also noted that, in my opinion, additional ventilation should be provided in the roof. (See Fig. 4.8 for typical ventilation tile.) My reasons for this recommendation were that lack of adequate ventilation is one of the causes of dry rot and bad ventilation can also, as previously mentioned, allow moisture vapour to build up in lofts.

After completing in the rear loft, I then made my way to the second trapdoor and entered the front loft. I was unable to fully inspect the second loft area, because it was dangerous to do so, but the conditions found were similar to the other loft area. (See report.) I then went down and prepared to inspect the rest of the property.

As the plumber and electrician were preparing their own reports on the individual services, I was able to concentrate on other matters but I still made a point on checking major items (e.g. sinks, baths, number of electrical points).

I found that the WC in the bathroom was faulty and asked the plumber to comment.

The wash basin was also cracked. I ran my hand under the wash basin and felt moisture. It was obvious that either the tap connectors were weeping or the crack was letting water. On inspecting the bathroom floor, I found evidence of leaking, but after removing an inspection board, satisfied myself that the floor and joists under were sound. The bath was a real antique. It was one of the old cast iron baths that had iron feet, in the shape of lions. Having no side panel, it was easy to check that there were no signs of rot on the floor underneath.

Other than the small isolated cracks to the ceiling areas, most of the bedrooms seemed in good order. (With the exception of the one with disrupted coving.)

My moisture meter confirmed leaks around the chimney breasts generally but this was only to be anticipated as the lead flashings had been found defective.

One of the rear bedrooms had an oriel window, one that pokes out from the face of the wall. In that room, I discovered additional signs that the flashings and soakers above were defective. The chimney breast, where it passed through a built-in cupboard, was damp. Downstairs the kitchen floor was made from red quarry tiles and was badly sloping which indicated some subsidence had taken place in the past. I tested the floor with a spirit level and this confirmed that the floor had settled. The skirtings were the old cement and sand type. I tested the walls with my moisture meter and the readings were very high. However, one has to be careful with moisture meters because the Protimeter type work by measuring the current flowing between the two 'fangs'.

I found a loose piece of wallpaper and carefully pulled it back. Underneath the paper, the previous owners had lined the walls with aluminium foil. Bearing in mind that aluminium is a good conductor of electricity and could be causing the moisture meter to give an incorrect reading, I had to check further.

Using aluminium foil is a common way of trying to prevent damp entering a building, but it is largely ineffective in practice. I immediately suspected the worst. I carefully eased aside some of the aluminium and found, as I expected, that the plaster underneath was damp and perished. The kitchen was a very damp room indeed, and it was probably only the heat from the old Aga stove that was preventing matters being worse than they were, by evaporating some of the excess moisture.

I made a note to advise that this floor be dug up and a new floor laid. I also noted that it should also be possible to salvage the old tiles if desired, and relay them as a decorative finish.

(I would refer the reader to later chapters for details regarding the old red tile floor finish.)

The kitchen units were in a terrible condition.

After completing the inspection in the rest of the ground floor rooms, I went down into the cellar, which proved to be very large and subdivided into three rooms; it had originally been ventilated to the open air by way of light wells. However, the light wells had been blocked off at ground level leaving the cellar virtually unventilated. Judging by the state of an old ledged braced door and of the windows to the redundant light wells, I suspected that an outbreak of dry rot might be affecting that area. When I opened the window, I was immediately greeted by the smell of decay and a fruiting body was evident.

One room was partially ventilated by the pipe that I'd found outside and there were signs that water had been pouring into the cellar for a long time. There was also evidence that at one time there had been airbricks in the walls but these had been systematically blocked up by the previous owners over the years and the whole cellar had a stagnant, musty feel to it. While poking around, I found a log that someone had brought down in the past as firewood. It had woodworm holes all over it and when I inspected timbers close by I found more woodworm holes. It was clear that the beetles had flown from the log and had infected other timberwork in the basement.

Once again, I made the recommendation that the house be inspected by timber decay specialists.

To some, this recommendation might seem something of a reneging tactic, but I don't think so. Remember, I had found the problem. It was now for the specialist to determine the solution, and the *cost*. The recommendation that I gave is a far cry from looking at the property for ten minutes and suggesting that dry rot and woodworm might be active. The client knew from my report that they *were* present and that treatment *was* needed.

Whilst in the cellar, I noted that all the brickwork in the walls showed signs of **efflorescence** and bad **spalling**. Efflorescence is a powdery residue which is sometimes seen on brickwork, particularly new brickwork, and is caused by water evaporating and depositing salts on surfaces. The thickness of the efflorescence in this cellar was about half an inch (12.5 mm) in places and indicated that the whole area was extremely damp and had been that way for a long time. The reason for saying this is because my assessment was that for years water had been coming through the wall and, as it evaporated, layer upon layer of efflorescence had been laid down. This damp level was reflected in the moisture levels recorded by the moisture meter in the cellar area. Once again, as with the roof, because of bad ventilation, the moisture in the surrounding air was dampening timbers.

Spalling is a technical term which means that the surfaces of brickwork are breaking up or laminating. The most usual cause of spalling is frost action on damp brickwork. The excess moisture in the bricks freezes and forces off the faces in exactly the same way that it breaks up clods of clay in a well-dug garden. Spalling can also be caused by expanding free lime in the bricks.

I noted that the heating pipes were all lagged but the lagging was asbestos and in places the outer canvas wrappings were broken which would obviously create a health hazard. Whilst I knew that the plumber would obviously comment on this point, I also noted that it might be best to have a specialist firm remove any old lagging. (Asbestos removal is now strictly governed – unlicensed removal and disposal of asbestos can lead to prosecution.)

Whilst in the cellar, I found an old cauldron which was obviously obsolete. I also found a sump and channels that radiated out from it and an electric pump. It was fairly obvious at this point, taking all things into account, that the proximity of the lake did affect the cellar and that in wet weather the cellar could flood. What is more it had been designed to deal with the flood water. However, a cellar that flooded was something that the client needed to know about.

I also noted that the floor had been strengthened in places. When the ends of the joists were inspected, I found them to be rotten. Some had had new timbers spliced in but the remainder needed attention.

From the cellar, I went into the garden. Looking around, I knew that the number of defects was going to be lengthy so I resorted to creating a simple list, starting with the greenhouse with 'bellied' walls and the front wall that had a 3 in (75 mm) lean on it.

As I was about to finish my survey, Mr Vernon reappeared. I had promised that I would give him a verbal report as the typed version would obviously take a little time to prepare despite the word processors that I had in my office.

As I ran through my list, Vernon asked me to let him have approximate costs of putting the property in order. I agreed to do so but stressed that they *would only be approximate* and that he would eventually have to obtain estimates from several builders if he purchased the house.

5 Tales of caution and neglect

Does a surveyor accept every job put his or her way?

Let us consider what the Institute of Building has to say on the matter. Under their rules of professional conduct, they state that 'A member shall only undertake an advisory service . . . in which he is competent and qualified by his education, training and experience'.

In this book, I have only considered houses of conventional construction (i.e. bricks and mortar). There are system built properties on the market which cannot be considered conventional and could very well lead a surveyor into dangerous waters.

I would cite one case as an example. I was recently asked to carry out a survey on an old army house.

The owner was complaining of cracks and bad slopes on the floors. When I drove around the estate (prior to seeing the owner), it was immediately apparent that the houses were of a non-conventional nature. It turned out that the house was constructed on a steel frame. (The owner exposed part of the first floor to show me the construction.)

I quickly came to the conclusion that my experience with this type of property was sadly lacking and that this was a job for a structural engineer.

In any case, as far as I could see, the only way of really ascertaining the cause of the problem would be by taking test bores below ground level and/or exposing parts of the structure.

Within ten minutes of being inside the property, I gave the owner the name of a suitably qualified consulting engineer whom I knew could deal with such matters.

So the answer to the question has to be 'no'. But I would like to pursue the issue a little further with a slightly differing slant. This time when asking the question, 'does a surveyor accept every job put his or her way', I want to attempt to examine motivations generally, and in particular the difference between any 'professional' and the average tradesman.

For the want of suitable workers, I intend to use bricklayers as a typical trade (I would stress that I have nothing against bricklayers!).

If you ask a jobbing bricklayer to build you a wall, he will usually quote a price for the job, and if you don't accept the price, and he has plenty of other work to do, then he will turn the job down without any qualms. In fact, with a self-employed tradesman, the more work around the higher the prices that he will usually charge. (Simple supply and demand.)

Does a professional automatically do this? I would suggest that in the short term the answer is 'no'. (In the long term the answer is 'yes'.)

Why is there a difference?

I know that I am making a sweeping statement and that there are always exceptions to every rule, but from what I know of the building industry, the average tradesman naturally reacts to demand. It is an almost instinctive reaction.

It is not uncommon for a typical bricklaying gang to move swiftly from one job to another within the space of a few weeks, and, in times of high demand, raise their price almost by the month for laying bricks. In the average tradesman's mind there is a direct link between doing the job and getting paid. They don't (in the main) lay bricks for the love of it. They do their job in order to earn money because it will buy what they need or want in life.

In my experience, the average professional's mind works slightly differently to that of the tradesman.

He or she is generally salaried and gets paid monthly and not weekly. The amount of work carried out, generally speaking, is not directly related to the effort put in. In other words, unlike the bricklayer, the professional is not paid by immediate results (e.g. as with a bricklayer, £X for every thousand bricks laid).

In addition, he/she tends to do what they do for a living, because the work is rewarding. They usually enjoy their job. For them, what they are doing is important and, generally speaking, not directly related to the monthly pay cheque.

In other words, they work with the desire to excel. All right, I hear you say, I know some very good tradesmen who get real satisfaction out of their work. As I said before, there are always exceptions to every rule.

What I am trying to indicate is that to the 'professional', the supply of the service is not directly related to the money at the end of the day. (Okay, so the more 'hard-nosed' amongst our ranks might react to demand more strongly than others, but I believe that in the main this is not the case.)

The effect of this difference is that it is possible for the professional to forget the realities of life and that he or she is no different from the tradesman. In the ultimate analysis, we all work to make a living.

This difference in attitude can create a problem for the professional.

I know when I first set up practice, I very rarely said 'no' when people asked for my help. I was offering a service but there were times when it was obvious that certain sections of my clientele didn't want to pay me for my labours.

After a while I began to analyse my motivation. For the first time in my life perhaps, I began to wake up to realities. There were jobs that were not worth having.

With regards to building surveying, this is particularly so when it comes to people who want a quick job done.

The favourite call is, 'I've got a crack/bulge in a wall . . . I want someone to have a quick look at it for me . . . It'll only take five minutes . . .' Or, 'I don't want a full survey . . . Just give it a quick whizz over, y'know what I mean?'

Sure I know what they mean.

Do you?

Translated, the first means . . . 'I'm trying to find someone who won't charge me anything for his/her time'.

The second means the same, except in this case you will be held responsible for ever and a day if the findings of your 'whizz over' miss some item or defect.

'Quickies' should be avoided at all costs. Legally, if a surveyor passes comment which is later proved wrong, even if he/she was not paid for advice, they can be sued. (If it can be proven that the incorrect advice was acted upon.)

I will mention one type of prospective client that is now becoming all too common.

Generally speaking, the 'prospective client' has just had a small extension built at the back of their house (say, a simple ten foot by ten foot flat roof extension).

So let's set the scene. Most of the interesting parts take place in the public bar at the White Lion Public House in Tipham. In this story there is quite a large cast. For ease I will name them thus:

(a) Mr Careful: He has decided that he wants an extension on his house.
(b) Mr Drawit: He draws plans as a sideline and volunteers his services.
(c) Mr Right: He is Mr Careful's drinking pal of many years' standing.
(d) Mr Knowitall: He is introduced to Careful once the problem of costs has reared its ugly head.
(e) Mr Slowbend: He is the local council Building Inspector.
(f) Mr Laywell: He is the builder.
(g) Mr Honest: He is a young, inexperienced surveyor who is finally called in by Careful when things start to go wrong.

And so our story starts to unfold.

Mr Careful is a lively fellow, and whilst in the White Lion, he bumps into a bunch of his mates and tells them that he is going to build an extension on the back of his house. It's the wife's idea. He can't see why she needs any more room in the kitchen. After all a kitchen which has a floor area six foot by eight foot ought to be good enough for anyone. Still, she's giving him 'earache', so he'll have to do something.

As if by magic, Careful is introduced to Mr Drawit, who claims that he is conversant with drawing house plans and as long as he is paid cash, no questions asked, he will sort something out for Careful.

Getting approval of the scheme takes longer than Mr Careful had hoped. (The local building inspector rejects Mr Drawit's plans on two occasions and the Planning Department insist that the length of the extension is reduced by 6 ft 6 in (2 m) because it falls outside their guidelines and cannot be granted Permitted Development status. (See *A Practical Guide to Single Storey House Extensions* by the same author for details of Permitted Development.)

Drawit tells Careful that Planning and Building Control are being difficult and there is nothing wrong with his plans. Although Careful seems to accept that the problems with Building Control and Planning are not of Drawit's making, he is not amused at the delays.

However, after a lot of arguing and amending, the scheme is finally approved.

Now, Mr Careful is happy and goes to the White Lion to celebrate. While he is there, Careful tells a friend of his, Mr Right, that Drawit has recommended that he obtains quotations from three local builders that he knows very well. According to Drawit, he has seen extensions that these three builders have constructed in the past. According to Drawit, they all turn out first class workmanship and therefore he has no hesitation recommending their services.

Right agrees with Drawit's suggestion. That way, Right argues, Mr Careful will be proud of the finished result.

Careful takes notice of this sensible advice, but like so many lay people, he has asked neighbours who have had house extensions built in the past (some as long as ten years previously) and has based his costing upon what he has been told.

When the quotations come back, Mr Careful screams 'How much!' and promptly falters.

It was a nasty shock, so bad in fact that he needs a drink. Whilst quenching his thirst in his local hostelry, he is introduced to Knowitall and tells the other man his troubles. He explains that the prices that he has been quoted are far too high. They are exactly three times the cost that he envisaged. Both men agree that builders, as a race, are among the lowest form of life known to man!

Knowitall then suggests that part of the problem could be that the builders that he went to are associates of Drawit, and who, according to Knowitall, may well give Drawit a 'backhander' for putting their names forward.

This suggestion starts to gnaw at Careful, and as we shall see later on, Careful and Drawit fall out with one another.

Knowitall goes on to suggest that Careful builds the extension himself.

He knows a gang of bricklayers who would do most of the work for him if he does the labouring. At first, Careful thinks that this is a good idea. Knowitall, however, tells him that it could well take him several months to complete his extension working at it part time. Careful seems to change his mind. He never has liked hard work.

Careful starts to change tack. If only he could find a cheap builder. Knowitall tells him that he must 'shop around'.

Never mind the quality, Knowitall says, go for the cheapest. After all, if the builder doesn't do a good job, you don't pay.

Careful weighs up the two arguments. Should he follow Mr Right's advice or should he do as Knowitall says? Finally, his mean streak wins the day.

Later on that week, he pulls out the *Yellow Pages*, the *Thomson Local*, the new business pages of the telephone directory and back copies of all the free local papers in the area.

He makes a list of 35 local builders and starts to ring around for prices. After many visits and after six weeks' hard interviewing, Mr Careful finally settles on a builder that has given him an absolute rock bottom price.

In his heart of hearts, he knows that Laywell Builders have made a mistake in their pricing. He suspects this because they are far below all the other quotes that he'd had, and after all, he's had 34 other quotes to compare it against.

Mr Careful's doubts return, so as always in time of stress, he pops along to the White Lion.

Mr Right advises him not to accept the cheapest price because when the builder realizes that he has made a mistake, the job will suffer.

Knowitall disagrees and tells Careful that he will be a fool if he lets this chance slip through his fingers. After all, if Laywell Builders have made a mistake, that's their problem, isn't it?

Mr Careful weighs it all up. He doesn't like making decisions at the best of times. Should he do what Mr Right suggests or should he follow Knowitall's advice?

Decisions, decisions!

Eventually, his mean streak wins the battle and the following day Careful sends a letter accepting Laywell's quote.

And why not?, you may say. It's about time the common or garden customer won for once. The answer is, despite what some people may tell you, in this life, you get what you pay for. In the main, if you pay peanuts then you will end up with a load of monkeys!

However, back to Mr Careful. He's been talking to Knowitall in the pub again. He says that his builder in his quote has allowed for facing bricks worth £300 per thousand. He says he's done this because Mr Drawit has not specified on his plans the exact type of brick to be used. All that Drawit has said is that the extension is to be built of facing bricks or specially selected commons to match the rest of the house. Careful complains that Drawit was going to obtain a sample brick for the Planning Department to approve, but he hasn't done so. Careful thinks that this is poor.

(Drawit is added to Knowitall's list of low life forms.)

What Careful omitted to tell Knowitall is that Drawit has not been paid for his plans.

(Remember how Knowitall planted the suggestion in Careful's mind about 'backhanders'. It was pure speculation on Knowitall's part, and in this case, totally untrue but Careful conveniently decided that it was so. To Careful's way of thinking, he doesn't see why Drawit should be paid twice for the same job. Besides, it took a long time to get approvals.)

(Naturally enough Drawit is not prepared to do any more work for Careful until he sees some money. In fact, Drawit is sufficiently displeased with Careful's attitude that he is considering taking him to the small claims court. But that is another story, except to say that as Drawit hasn't anything in writing to say that Careful owes him money, it probably won't be worth his trying.)

But back to the White Lion. Knowitall has a flash of inspiration, laughs and asks to see Laywell's quotation, then rubs his hands with glee and does a quick calculation on the back of a cigarette packet.

He works out that Laywell's have allowed for two thousand facing bricks in the extension. After a few further doodles, Knowitall sees a way of making some money for himself.

There is no way, Knowitall pontificates, that any builder will buy bricks at £300.00 per thousand. He tells Careful that he's being duped.

Knowitall then plays his joker. He tells Careful that, as the client, he should be able to choose the bricks. What is more, as the builder has set out the basic cost of the bricks that he intends to use, that Careful will be entitled to a credit if the prime cost sum of the bricks is not exceeded. *(No mention is made of the possibility of matters going the other way.)*

Careful hadn't thought of this. Now he has swallowed Knowitall's story, hook, line and sinker. Could this be a way of saving money?

Knowitall explains that Careful should tell the builder that he wants to see invoices to substantiate the basic cost and demand a credit if one is going.

Careful thanks Knowitall for his advice. Knowitall smiles and prepares to haul in his catch. Of course, he says almost dismissively, there is another way to save even more money. Now Careful is all ears.

Knowitall has a friend who can supply cheap bricks. He asks Careful if he knows what he means and then slides a finger down the side of his nose.

(Careful suspects, quite rightly, that they are about to fall off the back of a builder's merchants supply truck but as he doesn't know the laws regarding being in receipt of stolen property, it is a matter that doesn't worry him.)

It seems that Knowitall's associate can supply bricks at £50 per thousand (cash in hand of course). By so doing, Knowitall explains, Careful can legitimately take £600 off Laywell's price and save himself £500.

To be fair to Knowitall, he tells Careful that he will only be getting common bricks, and not facings, but then follows it up by extolling the virtues of the product.

Common bricks are what you want, Knowitall advises, why waste money on anything else? After all, no one will see the extension at the back of your house, will they?

The following day, Mr Careful, our money-conscious friend, tells Laywell that he wants to save money and tells him about the cheap bricks that he intends to buy. Laywell queries this but, being short of work, is bullied into giving the credit. Careful then verbally instructs Laywell Builders to use the common bricks which will be coming the next day.

Laywell meets the supply truck. The driver is hurriedly paid and promptly operates the tipping mechanism and is away in under two minutes flat, leaving a trail of chipped bricks deposited in an untidy strip in front of Careful's house.

Fred Laywell is not happy about what he has just seen and picks out a few samples from the debris to show Careful.

In Fred Laywell's words, 'The bricks are shaped like bananas . . . How do you expect us to do a good job with these?'

He then goes on to say that if this type of brick is used, that the extension will look hideous. And what about the chipped arrises (the edges of the bricks)?

Careful, who is now £100 poorer and still determined to save money, loses his temper and tells Laywell to get on with the job and stop quibbling. That night in bed, Mr Careful thinks over the happenings of the day. By a bit of smart negotiation, he has saved himself approximately £500 on his extension.

His common sense keeps telling him that something will go wrong but it is swiftly silenced by the more avaricious half of his mind. *It's only at the back of the house, isn't it? Anyway, a saving here and there, it all helps doesn't it? In any case, it will help to pay for a holiday, won't it?*

Laywell starts in earnest on site and digs the footings. Unfortunately, Mr Slowbend from the council says that he is still on filled ground and will have to dig the footings deeper. He'll be happy when Laywell hits firm clay.

Laywell does as instructed.

Eventually, he is six feet down and not three feet as shown on the plans.

Now Careful has been keeping an eye on developments and is anticipating that Laywell will ask for an extra. Careful decides that it is all a ploy to get the £500 back. That night, he takes a copy of the approved plans to the White Lion and he has a word with Knowitall.

Knowitall quickly notes that the plans state that the footings should be built to a depth agreed on site with Building Control.

He then explains to Careful, with thumbs shoved behind his braces, that, in his opinion, the contract with Laywell includes for foundations of any depth. It wouldn't matter if the builder had to dig down to Australia, Knowitall trills.

As soon as Laywell raises the subject, Careful chops the legs from under the other man, and then, two days later, with his £500 burning a hole in his pocket, goes on holiday, leaving the builder to his own devices. Anyway, to cut a long story short, Careful gets back from holiday two weeks later and is surprised and amazed that the job isn't finished.

Laywell Builders, by the way, stopped work because the drawing they'd been given didn't provide enough information and there was no one around to give them firm instructions.

That night, in a huff at not having his extension completed, and with his wife nagging him about the cement dust that has fallen in a thin film over the whole kitchen, Mr Careful inspects the half built shell and decides that he doesn't like the look of it. The brickwork is terrible. It has wide joints and all the brick arrises are chipped.

This upsets him no end and he rings Fred Laywell. Laywell reminds him that he was warned. This upsets Careful even more and he tells Laywell that he is fired.

Despite the fact that he has blamed Laywell, Careful realizes that Knowitall is also partially to blame. After all, Knowitall suggested the type of brick that should be used. But of course, now that the 'balloon has gone up', Knowitall has, as if by instinct, gone to ground.

Careful decides to get someone else in to look at his extension. So he rings another builder and tries to get him to give him a quote for putting the work right. What Careful did not appreciate was that most builders are too busy and far too astute to get involved in this sort of wrangle. Besides there is the builder's unwritten code of ethics to contend with.

Very few builders of any standing will stab a brother builder in the back. (And if they do verbally, they won't put it in writing.)

Eventually, Mr Careful complains to Slowbend in the Building Control Department.

Slowbend shrugs. It might look horrible, built in chipped common bricks, but it is structually sound and that is all he is concerned about.

By now, Mr Careful is at his wits end and decides to call in a surveyor.

Naturally, the surveyor who gets the job will be the cheapest.

Poor Mr Honest, the surveyor, 'draws the short straw' and meets Mr Careful at his house. Careful takes Mr Honest around the back of the house and shows him Laywell's work. Honest tut-tuts and agrees that it 'looks like a dog's hind leg'.

Obviously, as far as Mr Honest is concerned, Mr Careful has a genuine grievance. Careful tells Honest that he wants him to condemn the extension so that he can avoid paying the builder.

Honest goes away and starts to prepare his report based upon the facts as he sees them, but before he finishes his work, he happens to meet Mr Slowbend, the Building Inspector. Quite by chance, the subject of Mr Careful's extension is discussed. Slowbend doesn't know the full story but as builders sometimes seem to complain to everyone in sight when things don't go all their way, Slowbend knows something of the history.

In an unguarded moment, he tells Honest who the builder is and makes him aware that the client purchased the bricks.

Now Honest is troubled. If he had known from the start that Laywell's were involved, then he would have turned the job down. He was a fool not to have asked who the builder was in the first place.

He knows Laywell's too well and realizes that his loyalties are now badly split.

Unofficially, he telephones Laywell, and Laywell's side of the story comes out. It confirms what Slowbend has told him. Luckily, Laywell has had the common sense to write to Mr Careful and tell him that he cannot accept any liability for the appearance of the brickwork and that he had advised against the use of that particular type of brick. Laywell sends Honest a copy of the letter. Upon receipt of the letter, Mr Honest realizes that there are very few real 'defects'.

Honest has a problem. He has expended time on Careful's report but now he also knows that his client is not 'whiter than white' as he'd claimed. But was it his problem?

He could just issue his report, get paid and let events take their course. Had he not known about the bricks, then that is exactly what he would have done. But he does know!

After much more thought, Honest decides to telephone Careful and asks him for his side of the story. Did he buy the bricks? Did he receive a letter from Laywell's? Careful denies that he supplied anything, issued any instructions to Laywell's about bricks or received any letter from Laywell's. But it was sent recorded delivery, Honest reminds him. Careful mumbles something about his dog having eaten letters.

After a great deal more thought, Honest decides that it would be far better if he pulled out from the whole job and writes to Careful.

Upon receipt of Honest's letter, Mr Careful becomes livid with rage but undeterred! He starts to ring around for a surveyor who will give him the sort of report that he wants. There is no doubt that he will eventually get someone who will either not be as truthful, or as aware of the situation, as Mr Honest.

Does it sound like the sort of 'piggy in the middle job' that you would like?

But before we continue, and return to the original question posed, let us consider what we have learned so far. Let us consider Mr Honest's position.

There are lessons to be learned and questions to be asked that could possibly have several answers. For instance.

(a) If Honest had been a bit more worldly wise and more experienced, would he not have tried to obtain more detail from Careful before proceeding? Could he not have asked for the name of the builder when Mr Careful first telephoned him? If he had, like as not, he would have saved himself an unnecessary visit. No doubt he could have told Careful that he knew the builder concerned and that he might be faced with a conflict of interests.

(b) Should he have asked for some sort of deposit from Careful before starting work? Some solicitors do. What about a minimum call out fee payable in advance?

(c) Although I have presented this story as a more or less black and white case, what if Laywell's had *not* written to Careful confirming the instruction to use common bricks? Would that have altered Mr Honest's appraisal of the situation. Probably not in this case. We get the impression that he knew Fred Laywell very well and so would probably have declined to act anyway. It does prove though the value of builders confirming instructions given.

(d) Despite Laywell's dislike of the common bricks, could he have been expected to make a better job of the wall? Now that is a very subjective estimate, isn't it? In this case, we will never know.

There are some definite pointers to bear in mind though, and possibly some more food for thought. When it comes to house building in the UK, in my experience, everyone hopes that they will get a Rolls-Royce job but they only want to pay for a Mini Minor! The UK is becoming more litigation conscious and there are also large numbers of people who are ordering work and not paying for it. It is becoming an everyday occurrence. That is one of the reasons why so many decent small builders go into liquidation.

What we do know though is (returning to our story) that:

(a) Mr Honest has spent several wasted hours trying to sort out Careful's problems.
(b) Honest will only get paid if he sides with Careful. On the face of it, Careful sounds as if he is his own worst enemy.
(c) Although he has wasted time, Mr Honest preferred to keep his reputation. (And hopefully uphold the standards that surveyors everywhere should strive to maintain.)

Now we return to the original question. Should a surveyor accept every job?
The answer most certainly has to be 'no'.
A surveyor has to make a living and he/she is not in business to help every lame dog over stiles, no matter how much he/she would like to do

so. Neither should he/she devalue their profession by helping wrongdoers benefit from their devious schemes.

But how do you avoid getting involved with people like Mr Careful? I have hinted at the answer and, rightly or wrongly, find that it works well for me.

The answer is a minimum 'call out' fee.

The likes of Mr Careful don't like parting with money, and I find that if they know that they are going to have to pay something for your trouble, then they usually take their unwanted problems elsewhere.

And so, on to Ninetails Avenue:

Note
(See Fig. 5.1 for sketch of Ninetails Avenue. As previously indicated, Appendix B is based upon a genuine survey but names of clients, the location plan, addresses and the like have been amended and fictitious ones used to ensure anonimity of all parties. Sketches similar to those depicted in the figures were incorporated into the original report so that major defects could be located at a glance.)

The survey of the house in Ninetails Avenue is of interest for two reasons. The first because I nearly declined to act (for reasons that will become obvious later), and the second because this particular house had so many defects that I decided that I just had to include it in this book.

Ninetails Avenue seemed to have all the hallmarks of 'trouble' right from the start. I think it could well have been if I had not been cautious from the outset.

My involvement with 72 Ninetails Avenue started when I received a telephone call from Mr Bond. The telephone link was very bad and, sitting in my office, it seemed as if my prospective client was in some sort of echo chamber.

He told me that he had been recommended to me by a friend, but divulged no names, and asked me for an approximate quotation for surveying a property in Often Narrows. As I have said in previous chapters, I tend only to provide a 'budget' cost unless I can realistically assess the likely number of hours in a survey.

The conversation came to an abrupt end when the line was cut, and, as I had no way of contacting Mr Bond, I shrugged my shoulders and waited for him to ring back. He never did and for several days I thought no more of Mr Bond or his house in Often Narrows.

A week or so later, I was contacted by Mr Tree. He reminded me of my previous conversation with Mr Bond, who was apparently out of the country at that point in time. I quickly turned to the jottings that I had made in the notebook that I keep by the telephone and recalled my brief discussions with Mr Bond.

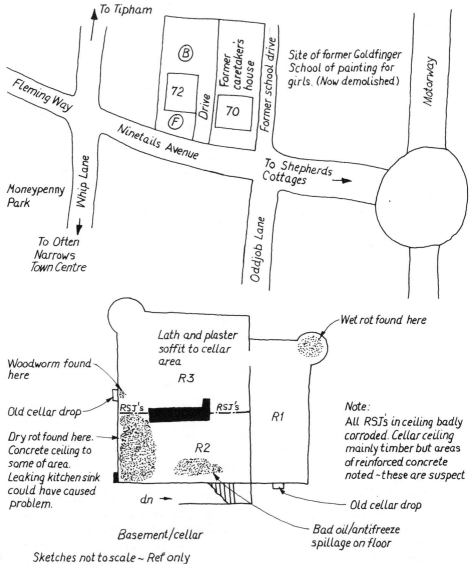

(a)   Sketches not to scale ~ Ref only

Fig. 5.1(a) and (b)   Sketch of 72 Ninetails Avenue.

Mr Tree informed me that he was acting on behalf of Mr Bond and asked me to carry out the survey as discussed.

I thought about the request for a few seconds and then asked who I would be working for, Mr Bond or Mr Tree?

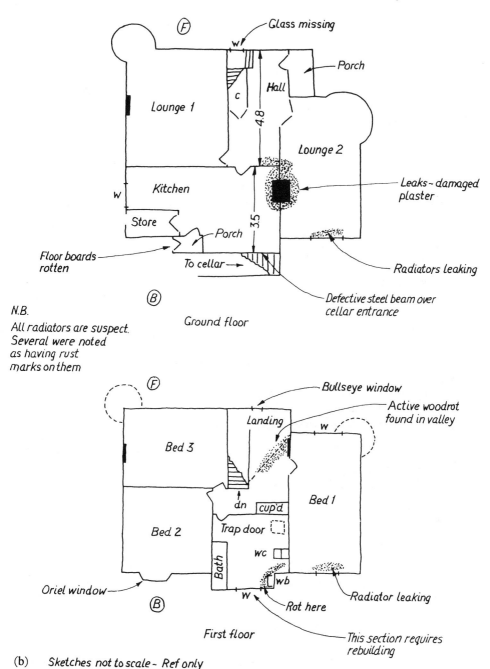

(b)   Sketches not to scale - Ref only

Fig. 5.1   Continued

Tree dodged around the issue but finally told me that Mr Bond was paying. I asked for an address and telephone number for Mr Bond. Tree didn't have either.

My reluctance to do business must have come across because Mr Tree started to change tack again and said that Mr Blofeld would pay if Mr Bond let me down and failed to meet his obligations.

I queried the connection between Bond and Blofeld. I knew a Bill Blofeld very well. Tree informed me that Bill Blofeld wasn't involved but his cousin Fred was. It all seemed a little suspect and I was beginning to look for ways of politely declining to act for anyone.

But Tree was persistent. In the end, I took Tree's number, told him that I would have to check my diary and said that I would ring him back.

I then spoke to Bill Blofeld. He told me that his cousin was a genuine fellow but he had no idea why he should be connected with either Tree or Bond and he said that he would speak to his cousin to see if he knew anything about the survey.

Shortly after the conversation, a fax appeared from Fred Blofeld but its wording was very vague.

I decided to accept but only on the condition that Fred Blofeld paid my account and I said so in the fax that I sent back. I also made a point of stating very clearly that I would expect immediate payment prior to issuing my report.

A return fax came within minutes accepting my terms and informing me that I would be acting jointly for Mr Blofeld and Mr Tree. Mr Tree, by the way, lived next door to the property in question and it was from him that I was instructed to obtain the keys.

Lesson number one. Always make sure who is responsible for paying your bill. If possible get it in writing.

Anyway, back to business. The house at Ninetails Avenue was condemned to death by bureaucracy. They killed it as surely as if they had taken a can of petrol and set it alight. However, I digress. Who am I to sit in judgement? They didn't break any laws.

As a purely technical exercise, the survey was very interesting.

In reality it was a 'one off'. It is doubtful if I will ever be asked to carry out a report like it again, but it is the occasional unusual job that helps to add a bit of interest to life, even if tinged with sadness at the probable loss of what had obviously been a very fine house.

The terms large and small are very subjective but I am sure that perhaps 80 per cent of the home hunting population of this country would give their eyeteeth to own a house like 72 Ninetails Avenue. Admittedly, the house has only three bedrooms but each one is almost double the size that you get in the usual modern house. In addition, the property possesses two large living rooms and each had feature windows (Fig. 5.1).

The overgrown rear garden would have taken two more detached houses in comfort and still have had space to spare. The rear of the property overlooked what had once been a school playing field.

No, in case you are wondering, I'm not into estate agency and I doubt if you would want to buy this house now. (Not unless you had a lot of money to spend.)

At the agreed time, I called at 70 Ninetails Avenue. Mr Tree didn't give me the keys immediately. Instead we talked. The story started to piece itself together. My caution had been justified. It turned out that the property was in some sort of joint ownership. I also gained the impression that the arrangement was highly unofficial and that the title deeds were held by Mr Bond.

Tree finally gave me the keys and I let myself in. I had been there about an hour when Fred Blofeld turned up. He came over and introduced himself. After the pleasantries were over, he deliberately took me around the outside of the property and started pointing out defects to me. (I had in fact noted most of them already but it was obvious that he wanted to ensure that nothing was missed.)

As the conversation proceeded, it became fairly obvious from things said, and the tone of Mr Blofeld's voice, that Blofeld and Tree wished to convince Mr Bond just how dilapidated the building really was. (And presumably how foolish Mr Bond would be if he dissolved the partnership.)

As the facts unfolded, I was thankful that I had ensured that my conditions of engagement clearly confirmed who was responsible for paying me for my services. There is nothing worse than a three corner fight. In such situations there is usually only one loser . . . and that is the weakest party . . . in this case myself.

Now as you may have realized, another lesson of this chapter is caution. Dismissing payments for the moment, there are obviously many areas where caution is required. Not only has the information provided in a report to be accurate and helpful, the prudent surveyor ensures that he/she records as many salient facts as possible within the report so that in the event of there being a query at some future date, there can be no doubt regarding the conditions of engagement, conditions met with on site or restrictions imposed.

The Ninetails Avenue report is also interesting, as I have said before, if viewed as a purely technical exercise. Figure 5.1 gives a fair indication of some of the major defects, the areas affected and the basic structure of the property. (These sketches were made in addition to the notes taken on a pocket tape recorder.)

As indicated above, I included a sketch similar to Fig. 5.1 in the report because I wanted to ensure that all parties could appreciate where and how extensive the problems were and to ensure that they could relate the written descriptions to exact locations.

NB The inclusion of sketches to assist understanding and reduce the length of a report is in my opinion a sensible approach, especially as modern photocopiers are nowadays almost universally available.

Returning to what I said about caution, you will note, when you read the report (Appendix B), that I included a preamble entitled 'The recent history of property'. I did this for two reasons.

The first was to ensure that information concerning the property was permanently recorded and that anyone reading the document (you can be certain that, despite the rider concerning ownership of the report, it will end up passing through a few hands) would realize very quickly that I was describing a house that had been neglected for a long time.

The second reason for including the preamble was because it reinforced paragraph five of the 'Instructions, limitations and general preamble'.

I wanted to ensure that anyone picking up the report would realize that I had been called in *after* the house had been purchased and not before.

As you will gather, the property in question had been under a **planning blight** for many years because there were several proposed routes for the future Often Narrows to M – link road; 72 Ninetails Avenue was one of 50 or more houses blighted in this way. (I don't know if the rest suffered a similar fate, but if I ever am asked to look at a house in the same area, I will consider the possibility more than just in passing.)

For those not familiar with the term 'planning blight', this is when a Council or Local Authority has proposals for an area which are still in the process of being developed. Because rumours spread quickly, properties in 'blighted' areas tend to lose value or in bad cases become valueless. In this case, 72 Ninetails Avenue would have been demolished if 'Route A' had been adopted. When 'Route C' was finally decided upon, the blight was lifted but the property had suffered during the intervening years. If you read the report thoroughly, it will become obvious that the house had originally belonged to Mr Bond's late father and was finally repurchased by Mr Bond (with the financial help of Tree and Blofeld).

As I have said before, long before I arrived on the scene, the three of them had bought the house, presumably with the intention of rectifying the defects and then placing the house back on the open market.

I also gained the impression (whether correctly or not) that now that the house was back in the family, Mr Bond resented the fact that Blofeld and Tree had an interest in what he saw as the family seat (dilapidated though it was).

Still, I am speculating and perhaps I shouldn't. So no doubt it is best that we don't dwell on this matter any further. Speculation is after all something to be avoided unless it happens to be on the stock exchange and you know what you are doing. So, without further diversion, I will proceed.

The house was unoccupied when I visited. It quickly became apparent that the whole place was a shambles. It was a house that was dying through lack of love. It had obviously not been painted for years and essential repairs had been ignored. Cupboards were stuffed with long abandoned clothing and furniture of all sorts cluttered rooms.

The cellar was the worst. (I found all sorts of strange items cluttering the cellar, ranging from an old set of stairs, which had woodworm, to dozens of packing cases.)

In order to make a systematic survey, under what were obviously going to be difficult circumstances, I decided to take a series of photographs of every elevation. I knew that these photographs would help with the survey and also act as a permanent record of site conditions at the time of survey.

NB I carry a Polaroid camera for this purpose. Although the films are expensive, the Polaroid has the advantage of providing 'instant' pictures. If one is defective another 'snap' can be quickly taken. The film of a conventional camera obviously needs developing and, if one shot is defective, a return visit is needed. Also you either waste a great deal of film or wait until you have used up a roll. This would obviously cause an unacceptable delay.

After taking the photographs, I made a rough sketch of the property as previously mentioned on all levels (Fig. 5.1(a) and (b)) and then referenced the photographs to the plan.

Coming back to the final report for a minute, as Mr Tree and Mr Blofeld knew the property well, I did not waste time providing dimensions and confirmed the fact under 'The property generally' by saying: 'As you are familiar with the property and area, we have not provided superfluous information such as the distance from amenities or room sizes.'

(I usually include this clause anyway and agree the same with my clients because estate agents usually provide this information. Some surveyors suggest that giving room sizes is mandatory but feedback from clients in my area indicates that they consider such information as 'padding', i.e. the surveyor has just put it in to justify a high fee – something which is usually not true in my experience – but Joe Public always suspects the worst despite the fact that surveys are usually very labour intensive. My attitude is if the client is happy and they are in agreement, then give them what they want. The only thing that I won't do is a 'quickie'.)

Neither did I describe the surrounding area. *(Except for the site of the former Goldfinger School of Painting for Girls . . . which was important because it was directly behind the property and seemed to be connected in some way to the domestic drainage system of the house. I also made a point of recording that the future of the old school site was still undecided.)*

Where dimensions were taken, it was for proving that the lower beams and walls were in the right position to adequately support the walls above.

As you will know from previous chapters, I like to work in a systematic manner, starting externally, going into the roof space next, going through the upper floor after that and then dealing with the ground floor.

However, in the case of Ninetails Avenue, after carrying out my external survey, I then went into the cellar/basement. The reason for this was that during my initial sketching phase, I had noted a truly massive outbreak of dry rot in the basement area and it seemed rather pointless not investigating this first and then applying the knowledge to the floors above. The reason for this break with normal procedure becomes obvious when one understands the nature of dry rot (see later chapters).

The outbreak in the cellar/basement had obviously gained a strong hold and as dry rot can force its way behind plaster and through brickwork, it seemed sensible to pursue it from its source.

If you read my report, you will note that I inserted two sections into my standard report that spelt out in words of one syllable the effects of wet rot, dry rot and woodworm and then followed it up with a section on rising damp. In this case, I considered that it was essential to leave my clients in no doubt that the house was in a bad condition.

You will also note that I made it quite clear that I suspected that the outbreak in the cellar was 'only the tip of the iceberg'. This, in my opinion, was not an unreasonable statement to make considering the numbers of burst pipes and radiators in the property and the fact that the floors, walls and ceilings had a high timber content.

The walls and ceilings were constructed of lath and plaster, a system used some years back before plasterboard was widely used in the UK. It comprises small strips of wood (the laths) which are nailed to the walls and ceilings. Plaster was then spread over the laths. The action of the plasterer spreading his plaster forced some of it through the laths, so creating plaster 'dovetails'. (In some cases, hair was mixed into the plaster as 'reinforcement'.)

In a house like Ninetails Avenue, all the dry rot fungus had to do was to follow the wood in the ceilings and partitions until it found the floor joists, ceiling joists and rafters. Not to have warned my clients of such a possibility would have been very wrong.

You will also note that I advised an inspection by an eradication company. As I have explained before, this in my opinion is fair comment, not a 'dodge'. I had found the dry rot/wet rot and I had to the best of my ability indicated where I had found it but tracing the true extent of the outbreak was well beyond my brief. To do this would involve cutting inspection holes in plastered walls and lifting large numbers of floorboards.

Now let us consider the cellar (Fig. 5.1). For convenience I divided the cellar up into three rooms (R1, R2 and R3). The cellar generally was in bad condition, even ignoring the dry rot.

The first sight to greet me on entering the cellar area was a corroded rolled steel joist (RSJ) that was the main support to the wall over the cellar steps. Perhaps I had best explain. It is unusual to leave steel beams unprotected (or any steel for that matter). Beams are usually encased in some way (e.g. with concrete, plasterboard, etc., in order to prevent them rusting or being adversely affected by fire, in the event of a conflagration).

However, the houses at Ninetails Avenue seemed to have been constructed with the supporting RSJs open to the elements and with very little protective paint apparent.

As a consequence, the beam over the external steps to the cellar was a perfect example of the effects of rust expansion. At the time of inspection, both flanges of the beam were about three times their original thickness and the thrust had disrupted the brickwork to the bathroom wall above.

Twenty-five mm cracks in the walls were apparent and it was fairly obvious that the only way to repair the bathroom 'outrigger' was by taking down the affected brickwork, replacing the RSJ and then rebuilding the wall. Later on in the survey I discovered that the internal walls to the house were also built off other unprotected steelwork and I advised my clients to consider consulting a structural engineer if there was any doubt concerning the strength of the beams. I also warned that the beams should be fire clad or painted with an approved **intumescent paint**, these paints protect steel against fire. (See report for further details.)

The second sight to leap at me was a dark sticky patch on the cellar floor. It turned out that this was a spillage of car antifreeze. Once I had discovered what this was, I was less concerned. I would suggest that you now turn to the report. (See Appendix B at the back of the book and note the full comments made.)

After carrying out the cellar survey, I went back to my usual routine and went up into the roof space. From the outside, I had noted that there were one or two slipped slates but nothing had prepared me for the stench as I opened the trapdoor to the loft. The whole roof space was alive with flies.

Within seconds the lead lamp was covered in them. I made a valiant attempt to check out the loft but after 20 minutes I retired. Not only was it difficult to see anything, someone in the dim and distant past had in places laid additional glass fibre quilt across the ceiling joists (instead of between the joists) which made progress very dangerous.

NB If you cannot see the joists, it is easy to put your foot through the ceiling.

I noted all these factors within the body of the report for future reference.

You will note from the report that there was one section of roof into which I could not gain access. This too was recorded. Note that I also hazarded a guess at the cause of the fly infestation (i.e. dead birds . . . The implications being that the flies (or rather their grubs) had been living off the carcasses). I also warned against overloading the existing rafters in the event of the roof being recovered. This can happen when a lightweight roofing is replaced with a heavier substitute without checking that the roof members are adequate.

After completing what I could in the roof, I then checked out the first floor and ground floor, bearing in mind what I had seen in the cellar.

All the rooms reflected that the house had not been occupied for years. Wallpaper was hanging limply and it was obvious from visual inspection that the majority of the plasterwork on the ground floor walls had perished. The kitchen was particularly bad.

There was evidence that the central heating radiators had perished, and I had also had confirmation from Mr Tree that there had been several instances of where the whole house had frozen during winter months. He also informed me that the leaks to the plumbing systems had only been discovered when water was seen cascading from under one of the external doors.

All in all, the house was in a sorry state. In areas where the water had been running freely (Fig. 5.1), in the hall, kitchen and lounge 2, moss and mildew were apparent. The water problem was also being added to by leaks over the bays in both lounges and missing glass in several windows.

Evidence of rats was found in the kitchen and a further wood rot out-break was found in a valley beam above the landing.

Once I had completed inside, I went into the garden areas. As you may have gathered, during Mr Blofeld's visit I had already noted many defects but I still went around carefully adding to the list.

The main problems seemed to be tree damage and leaking drains. (*The two were probably connected.*) The tree roots had probably forced their way into the drains which then caused the paths to subside as the ground support was washed away. Note that I advised that no tree be felled until it had been checked to ensure that it wasn't on a tree preservation list.

My conclusion speaks for itself. I declined to provide any budget cost for remedial works. In this particular case, it would probably have taken far longer to assess the true cost of rectifying the damage to the building than preparing the report. (That is, if the job was done properly and in conjunction with the relevant specialists.)

In case you are wondering, 72 Ninetails Avenue is still slowly dying because of lack of love.

As far as I am aware, the issue of the old school grounds has now been resolved. It is going to be a park. However, the owners of Ninetails Avenue seem to have lost interest in the property.

Perhaps one day it will be restored to its former glory.

(For full report see Appendix B: Ninetails Avenue report)

6   Botched alterations for resale

(For full report see Appendix C: Acacia Avenue report)

My involvement at Acacia Avenue came about in a roundabout way. A friend recommended my services and, as a consequence, Mrs Mayo asked me to do the survey. As the house concerned was somewhat out of my normal area, I was rather reluctant to take it on but was persuaded to do so.

I gained the impression that Mrs Mayo had no doubts about the property and intimated that it would take a lot to dissuade her from buying it but that her husband wanted a survey.

Once contact had been made, I treated Mrs Mayo no differently from any other customer. I sent her my standard letter and one to the present owner confirming time, date and approximate length of time that the survey would take. As I have just said, Mrs Mayo was keen not to 'lose' the property and wanted the results of my survey very quickly.

In order to speed matters up, I agreed to give her a verbal report and then confirm a few days later with my written findings, stressing however that the written report had to be the binding one. I also made it quite clear that the decision to buy, or not to buy, was hers. I merely gave the facts, and she decided, that was how I saw my position.

Not knowing how long it would take to reach the far side of Shepherd's Cottages, I set off early and consequently arrived ahead of schedule. On first sight, the property looked a good buy and I could understand why Mrs Mayo had fallen in love with the place.

Number 3 Acacia Avenue was a two storey semi with a spar dashed frontage. (The sides and rear were just plain render.) The first six or so courses of brickwork above and below the DPC were facing brick. The windows generally appeared to be stained single glazed hardwood and had obviously been installed recently. (Upon closer inspection, the windows were found to be softwood.)

The roof was pitched, covered with a smallish tile and had a secret gutter where it met number 1. I made a note to give the secret gutter special attention because, as usual, that was one area where I anticipated finding problems.

As I couldn't see anyone in the front rooms of the house, and as I try to adopt a rather Prussian attitude to time, being neither too early nor too late for an appointment, I didn't go straight to the front door and announce myself.

Instead, I studied the house for some time making sketches and taking photographs for future reference as described in previous chapters.

From what I'd seen, other than the fact that the adjoining semi (No. 1) had obviously not been looked after and would have to be taken into consideration, Number 3 Acacia Avenue looked in excellent condition.

Once again, from that position, other than the odd defect, the house looked good. It was obvious that in the recent past someone had spent a fair amount of time, effort and money on the place.

Whilst not lowering my guard for a moment, I was looking forward to being able to issue a very positive report.

I slipped the camera back in my tool bag, pocketed the photographs and then went to the front door and knocked. A moment or two later, Mrs Smith, the current owner, let me in. I asked her if she had received my letter. She confirmed that she had and that she would be at home during most of the survey but that her husband would be taking over from her in about an hour and a half. (I had advised the Smiths that I expected to be in/around the house for between 3 to 4 hours.)

Whilst talking to the owner, I'd been looking around the place. The decoration was superb and the floors were well carpeted. Why did the Smiths want to move? I asked the question and then hurriedly explained that I was very impressed and couldn't understand anyone wishing to leave. (Sometimes one has to apply flattery with a trowel.)

Mrs Smith informed me that they were only leaving reluctantly and that they had lived there only a year.

(Later on in the survey, when Mr Smith came home, I discovered that they were selling the house because they could not afford the mortgage repayments any more. He told me that he was now on short time working. Apparently he worked at a local factory but because of 'rationalization' the plant was closing. It was a common story in that area.)

Apparently the house had been owned by Jake Conroy, a local builder. According to Mrs Smith, Conroy had done everything to the house from painting, new windows and electrical rewire. I complimented the workmanship. It certainly looked good.

As I did not know the area very well, I took the opportunity of picking Mrs Smith's brains a little further and asked her if she knew how long the house had been built and if she could tell me when the extension had been constructed. (I had no estate agents details.)

I gained the impression that the house had been built just after the last war and that the rear side extension had been built by Conroy. From what Mrs Smith said, I gathered that Conroy had sold them the property only about twelve months previously.

The Smiths had been so impressed with the standard of decoration and the work done to the house that they had never bothered to have an inspection carried out by an independent surveyor.

I asked her if she knew of any defects. (It might seem a daft question but most people realize that if they are untruthful and are discovered to have told a blatant lie that it undermines their credibility.)

Mrs Smith said that as far as she knew the house was perfect. I asked her about possible rewires, new damp proof courses (DPC) and the like. She told me that a new damp proof course had been installed by Conroy, and he had done all the modernization that could be seen. I asked her if she had a transferable guarantee for the damp proof course. She gave me a rueful smile and said that they had accepted Mr Conroy's word that he'd installed a new damp proof course. No guarantee existed. She then, half with a smile, asked me not to find anything wrong. Please!

I made a mental note to check the DPC carefully.

At that point, I decided that I had learned as much as I could about the property from the owner and explained that I intended to check the exterior, the roof, DPC, floors, etc.

Mrs Smith accepted this philosophically and let me loose. I then went to unload the remainder of my gear from the car. As there was no access around the house, owing to the fact that a side extension now straddled an area that had once been a driveway, I decided to deviate from my normal procedure and complete everything on the front and side first, including external areas, as to do otherwise would have meant tramping dirt into the house on several occasions.

It would also mean that I would have had to move my equipment a great deal, and as I carry all the tools indicated in Chapter 18, and a few more besides, it is not a task to be taken lightly.

I then started to go back over my preliminary inspections. I was able to see the main front and side roofs very clearly from ground level and noted several chipped, slipped and missing tiles. These I duly recorded, slope by slope. I then inspected the front chimney. This seemed to have been well pointed, but upon closer inspection, using the binoculars, it was obvious that whoever had done the pointing had basically done a 'smear job' and that the top two or three courses had not been raked out at all.

The flaunching on the top of the stack also looked as if it required renewal. I next tackled the gutters and downpipes.

I had already noted that most of the high level gutters had been replaced in PVC but the downpipes were still the original cast iron system.

Using the 'mirror on a stalk' that I carry, I checked the pipes at the back. In several locations, the backs of the pipes were badly rusted. Rust was also appearing through the new paintwork on the fronts of the pipes. It was obvious that all the downpipes needed renewing and had probably been painted only as a 'cosmetic' exercise.

At the front, one small length of cast iron gutter still remained. Quite naturally it was also in bad condition and, instead of having purpose made stopped ends fitted, someone had placed curved pieces of wood at each end. As light could be seen at the joints, it was obvious that in wet weather the gutter would undoubtedly leak. I then tested the fascia and soffit boarding with a sharp probe and my moisture meter. The wood seemed sound.

I moved to another location and repeated the tests in that area. Upon coming down off the ladder, I went back to my tool box that I had placed in the porch to get some other equipment.

It was then that I noticed that some rendering, on an inner 'wing' wall inside the porch, had fallen off.

I inspected the area carefully and noted that the rendering inside the porch was generally in poor condition and that bulges and cracks were noticeable.

This immediately made me suspect the rest of the rendering. When I started to tap the render on the side wall, it rang 'hollow'. I then started to look for cracks and bulges.

The wall had recently been painted with an external stone paint, but under close examination extensive patching was noticeable. There was also no **bellmouth** where the render met the brickwork (see Appendix C, 'External render and spar dash areas').

It was obvious that the render was not in the best of conditions and, as cracked render can trap water instead of keeping it out of a house, I duly noted the possible defect.

As I have said before, the front walls were spar rendered but I found myself considering the possibility that the spar had been applied only to disguise defects. I began to examine the front of the house very carefully.

On the path at the front of the house there were small piles of spar that had become loose and had fallen off. There were also areas of wall that were already becoming 'bald'. I spent several minutes examining the spar dashed areas. After a while, I came to the junction with number 1. There was a cast iron downpipe at that point which drained the intersection of the roofs between 1 and 3.

It was obvious, by comparing the new dash with the adjacent house, that whoever had re-dashed number 3 had not bothered to remove the old render, but had merely gone over it with a thin layer of adhesive and then tossed on the new spar. The thing that gave it away was the slight step in the surfaces.

I came to the conclusion that although the new spar looked nice, given a few years it would probably all drop off leaving the old cracked rendering below on view. I repositioned my ladder and examined the spar higher up. The story was the same.

Whilst up the ladder I also inspected one of the new windows on the first floor. It was stained softwood but it was obvious that only one thin coat of stain had been applied and that the timbers were virtually unprotected from the weather. It was far from being a recommendation of Conroy's work.

By this time, my suspicions were well and truly aroused. The house had been newly spar dashed, the windows had been replaced and there was a new extension at the rear. Had Conroy bought the house very cheaply and then done the minimum amount of work on it to sell it off for as much as the market would stand?

Were there defects that were being hidden?

After a while, I decided to go next door to number 1. It was obviously unoccupied because the weeds in the front garden were nearly four feet tall and I suspected that, until recently, number 3 had been in a similar condition.

I paid particular attention to the brick joists which had not been re-pointed like those on number 3 because I suspected that there might be **cavity wall tie corrosion** apparent. If cavity ties in a property have reached a bad state of decay, the rust expands quite considerably and this, in turn, 'lifts' the bricks in regular bands all around the house. There was no evidence of any cavity 'lift' at number 1. (See later section for details of this defect; upon taking measurements, I later came to the conclusion that the walls of the house were solid construction and not cavity.) After looking around number 1 Acacia Avenue for a few minutes, I returned to number 3 and then inspected the DPC. There was one apparent at 150 mm above ground level but moss bridged it in places.

This immediately made me suspect the effectiveness of the DPC despite the fact that Mrs Smith had said that it had been replaced. I tested above and below the DPC with my moisture meter and found that the readings were virtually identical. (One has to be cautious using a moisture meter externally – heavy rainfall in the recent past can give misleading information – but it had been very dry weather for over a week.)

I then started looking for signs that an injection DPC had been installed.

Normally this operation is carried out from the outside and small holes can be seen where the fluid has been injected, but in this case no holes were apparent. So, unless Conroy had done the injection work from inside, no injection course had been installed. If this was the case, then Conroy had misled the current owners.

I recalled that Mr Mayo had also been concerned about the DPC. I made a mental note to check out very carefully the moisture levels internally.

After quickly checking the front garden and fences, I then lugged my equipment through the house to the rear, took a photograph and then repeated the process that I'd carried out at the front. To be honest, the rear was a repeat of the front and side, so I won't bore you by running through it again.

The rear completed, I went to the loft. After plugging in the lead lamp at a convenient socket, I hauled myself up through a very small access hatch and into the loft space. The loft was of the conventional type, with rafters and purlins. I checked the sizes and spacings so that when I returned to the office I could check them against my Building Regulations tables. (It is advisable to check whatever you can even though it is fairly obvious that the roof works. If it was drastically understrength Nature would have destroyed it long ago.)

The floor of the loft was unboarded and had approximately 25 mm of glassfibre insulation laid out. The underside of the roof was the old-fashioned render work, but it soon became obvious from tests taken with the moisture meter that the roof was very adequately holding its own.

Once the general inspection was completed, I wormed my way towards the valley gutter and the secret gutter (Figs 11.25 and 11.26), fully antici-pating to find damp. The house defeated me. The valley and secret gutter boards proved dry. (Within the 'green' zone on the moisture meter.)

Despite the fact the roof proved to be in good condition, I was by now highly suspicious of the house.

Once I returned to the ground floor, I started to take a series of damp tests on the skirting boards and on the walls directly above them. Every reading was in the 'red zone', meaning that timber decay was inevitable. (It could have already started.)

I then rolled back some of the carpets (where I could). I began to take an interest in the floor surface itself.

It was a cement and sand screed and it seemed new. All the damp readings on all the floors were high. It was obvious that the screed had been laid very recently and that it had probably been laid over an old floor that had no damp proof membrane in it.

Whilst this is not necessarily serious because concrete cannot decay like wood, if timber comes into contact with the damp floor, it will start to absorb moisture.

However, whilst I was checking the skirtings, I had noted that about 50 mm of the skirting was buried in the screed and, as I have just indicated, timbers will absorb water from damp surfaces very quickly. Bottoms of skirtings should not therefore be buried in damp screeds.

As far as I was concerned, I had that last piece of evidence that proved conclusively that the house had been 'bodged' for resale and I had no intention of pulling any punches in my report.

7 Further work will be required

(For full report see Appendix D: Hinter Lane report)

As I have outlined the basic method of carrying out a systematic survey in the previous chapters, I intend to cut the description very short in this case, but would stress that I followed the same method while going around Hinter Lane. All that I intend to do is give a brief background to the report.

In order to make matters easy I list the main parties:

(a) Mr Dove who wanted to buy 19 Hinter Lane, which was a small dilapidated bungalow on the top of Magpie Hill and very close to the Green Belt.

(b) Mr Hanson owned 19 Hinter Lane. The bungalow was left to him when his mother died and had been unoccupied for several years. Prior to entering into negotiations with Mr Dove, Hanson claimed to have obtained Outline Planning permission to build another house in the grounds of 19 Hinter Lane.

(c) Blackbird Damp Proofing (Often Narrows) Ltd were employed by Mr Hanson to rectify dry rot and timber pests that had been discovered in the property. The remedial work had already been carried out when Mr Dove made Hanson the offer.

At the time of my survey, I had been given a copy of Blackbird Damp's schedule of remedial works and based my report upon it. When I came to prepare the report, for simplicity I bound copies of this and other documents into the back of my report for reference purposes. (*The other reports were only copied after I had obtained permission so to do. These reports have not been reproduced.*)

Now down to details.

I had done a great deal of work for John Dove over the years, so it was natural that he should come to me when he was intent on buying a house.

When he arrived at my office, he showed me a series of photographs, Ordnance Survey sheets covering the area and several reports prepared by Blackbird Damp Proofing (Often Narrows) Ltd. They were the usual sort of document produced by a remedial company and guaranteed their works for 30 years subject to other works being carried out (these are listed later in the chapter).

I quickly read Blackbird Damp's report and formed a not too favourable impression of the property but Mr Dove explained that the remedial works listed had already been carried out. Mr Hanson had assured him of that.

I quickly read over the specimen guarantee because small firms of dry rot and rising damp specialists have a habit of going out of business, and what is the use of a guarantee from a defunct company?

However, it soon became apparent that the firm had the backing of a well known chemical manufacturer of some years' standing, so I accepted it as being of some worth (i.e. if the treatment company went out of business then the fluid manufacturer would automatically take over the guarantees and pay for any defective work).

I then queried items that were listed by Blackbird Damp that did not form part of their contract with Mr Hanson, namely:

(a) Renewing plasterwork where injection damp course had been installed.
    (*Without this work being done the injection guarantee was void.*)
(b) Renewing the roof coverings.
(c) Renewing the gutters and downpipes.
(d) Repointing all external walls.
(e) Providing ventilation to the floors, walls and the roof void.

I asked Dove if Hanson had had this work done. He was unsure but thought that he had.

Despite Dove's assurances, the bungalow still sounded and looked a wreck to me, but Dove seemed determined to go ahead, subject to my report.

As Blackbird Damp had given a full 30-year guarantee on woodwork and rot, I was instructed to concentrate on the structure.

I pointed out to Dove that Outline Approval didn't mean that he could automatically build on the site. He would still have to submit detailed proposals to the council and judging by the Ordnance sheet the piece of land concerned did not seem very large and any new property built there would have to be very small. I also expressed the view that if the site was of great value, it was unlikely that a man like Mr Hanson would be packaging it off in the way he was. Dove confessed that he too had misgivings.

He asked me to check with the Local Authority and make sure that the garden of the bungalow had received planning approval for another house as Mr Hanson claimed.

Mr Dove then went on to tell me that he didn't want me to comment on the old 'outriggers' at the back of the bungalow as he intended to demolish them anyway.

As the conversation developed I became more and more uneasy about what he was proposing.

I knew that he had spent years converting and enlarging his present home (*I had prepared four sets of plans for various extensions over the years*) but now he wanted to sell his house and end up with a property that in my opinion was far inferior in terms of accommodation and something that by his own admission was going to be uninhabitable for some time.

He talked about buying a caravan as a temporary measure and siting it in the grounds whilst he put the bungalow to rights. As Dove was then in his mid-fifties it seemed some undertaking to me.

Normally I don't challenge a client's intentions. It's not my place to do so, but knowing Dove as well as I did I asked the obvious question. Why did he want to move?

Dove explained that his wife liked the view from the top of Magpie Hill and that she had fallen in love with the bungalow. It was obvious that she was the prime mover in the matter.

It was also obvious that Dove himself was not so sure, otherwise we would not have been having the conversation in the first place.

(*When you have spent nearly seven years of personal back-breaking effort extending and altering a house, and had just got the place as you wanted it, as Dove had, then there have to be misgivings about starting the process off all over again.*)

I made arrangements with Hanson and a few days later went out with Mr Dove to give the property a full survey.

Loaded up with my ladders and my full survey kit, I went to Hinter Lane and found that the photographs had not lied.

Mr Hanson, the property owner, was there and for over 20 minutes waxed lyrical about the bungalow. I finally managed to disengage myself and was pleased to find that Blackbird Damp had left several sections of floorboards up. To be honest, I'd been itching to have a look at what had been done below the suspended timber floor and to get up into the roof space ever since Mr Dove had contacted me.

I turned on the mains power, donned my overalls, and then plugged in the lead lamp that I carry. There was space of about 2 ft 6 in (750 mm) below the timber floor so I slid my legs down the access hole and prepared myself for a very awkward crawl.

Dove appeared at the top of the hole almost as soon as I was down, asked me to wait for him and then clambered down behind me. He obviously wanted to see what I found.

As I moved beneath the floor I was immediately conscious of the smell of fresh earth. When I looked down, I realized that, like several older properties that I'd seen in the area, it had no subfloor. All that was below the timbers was soil. (*In the later chapters there are details of modern and older forms of construction.*)

The next surprise was to find fresh concrete in the floor space which did not seem to fit in with the age of the property and I quickly came to the conclusion that, at sometime in the recent past, the house had been underpinned.

(*Mr Hanson denied all knowledge of this work. Underpinning, by the way, is a process whereby builders tunnel under the old, defective foundations and replace them with new footings.*)

Armed with my lead lamp and moisture meter I started to work my way further under the building with Dove following closely behind. There were old bricks batts and bits of timber lying around everywhere. I made a note on my tape recorder that the subfloor area needed a thorough cleaning out.

As we went deeper I found a sleeper wall had been damaged by the treatment specialists and it also became obvious that the whole of the front and rear of the house had been underpinned.

The only reason for underpinning work (which is a very expensive operation to carry out) is that the existing footings had failed. When I inspected the rear walls from the outside, there were signs of subsidence to the bay windows and the walls themselves.

You will note from the report that I checked with Building Control and was told that they had no records of any underpinning work. In other words, in all probability the work had been carried out illegally and it was unlikely that it had been supervised as it should have been.

(*One of the things that property owners forget when they carry out works to their houses without obtaining the necessary approvals is that, sooner or later, someone will start asking questions. It could be a solicitor or a surveyor when the house goes on the market. If that person cannot provide evidence that the works have been supervised by Building Control and/or approved by the Planning Department, then resale value can be badly affected.*)

The next point of real interest was in the kitchen. If you recall, Blackbird Damp said that they had treated the property and that it was now free from woodrot. Within five minutes, I had found *Coniophora cerebella* (wet rot) behind a loose skirting board in the kitchen and obvious signs of woodrot elsewhere. It was a black mark as far as I was concerned. What else had they missed?

I would draw your attention to the section of my report entitled 'WC, bathroom and pantry'. At the time of my inspection, there was no bath in the 'bathroom' and I was unsure how one could ever have fitted into such a small space. I recommended that these rooms be combined.

The drainage system was also a mystery. You will note from the report that the only manholes and drains found were within inches of the surface of the ground and had saplings growing out of them.

I would now suggest that the reader studies the full report in Appendix D.

Did Dove buy the house? No, he didn't and I'm rather pleased. Despite the fact that his wife tended to get her every whim supplied, in this case, I think that Dove realized that Hanson had 'seen him coming' and was hoping to 'offload' a problem property onto someone else. In any case, as I said before, he'd spent over seven years, to my knowledge, altering and extending his old house, and I think that he decided that at his time of life, it was time to quit while he was ahead.

8 Concealing the evidence

(For full report see Appendix E: Fern Lea report)

As in earlier chapters I have described typical surveys in detail, I only intend to use this part of the book to highlight the reason why a survey is so important to any prospective home owner.

My interest in Fern Lea started when Mr Brown rang my office and asked if he could make an appointment to see me.

He made no mention of a Building Survey. The following day, at the appointed time, Mr Brown arrived in my office.

He quickly explained that since his wife had passed away, his present house was too difficult to maintain. His priority in life therefore was to buy a suitable bungalow in the area; one that would be easily maintained.

We talked for over half an hour, and then Mr Brown must have decided that he was going to engage me, because he gave me an estate agent's sheet. I made arrangements to visit the property and visited two days later.

Fern Lea was a small detached bungalow and on first sight looked superb. It had obviously been freshly decorated and the gardens were well tended. I usually reckon that I can find at least five defects as I walk down the drive but Fern Lea on first sight looked in excellent condition.

The owner, Mr Sly, met me at the door almost as soon as I rang the bell and invited me in. He'd obviously been awaiting my visit with bated breath.

He seemed a dapper man, and I imagined him being part of the lost empire! As I entered the porch, I immediately began to feel uneasy. The floor felt uneven but I didn't comment. Sly took me through a long hall and then into the front living room and then asked me to sit down. His wife was also in the room quietly knitting. Sly told me that before I started the survey, he wanted to have a chat with me. It was somewhat unusual but I consented. Did I want a cup of tea, he enquired? I accepted, even though it was taking up some of my time.

As I have already stressed, one has to bear in mind that when doing a survey, you are dealing with the public and it doesn't do anyone any good to be surly. Why antagonize the current owners by being churlish? In any case, the Slys might after all provide one or two snippets of useful information.

He sat me in an easy chair and then proceeded to tell me that the bungalow had been built in 1931–2 and had been constructed by Fred Bloggs. (I think I was meant to be impressed.)

Over tea, Sly assured me that the house had been surveyed prior to being put on the market and that the previous surveyor had given the place a clean bill of health, so he didn't expect me to find anything wrong. He asked me to bear all this in mind whilst I was looking around. His bungalow was, he insisted, in perfect condition, and that his wife and himself had only put it on the market very reluctantly. Very nice, but all the time that I was sitting there my shoe was worrying a lump under the carpet.

I asked if it was possible to see a copy of the other surveyor's report, just for interest. This seemed to surprise Sly for a moment. He made a tentative attempt to poke in a cabinet but then decided that it couldn't be found. All the time that this was going on, Mrs Sly's knitting speed increased alarmingly. I told them that it didn't matter and took a few sips of tea.

I extricated myself from the tea ceremony shortly afterwards and humped all my equipment in from of the car. All the time, I was conscious that something seemed wrong with the floors. I say 'seemed' because the house was fully carpeted and it was not possible to actually see the construction underneath. I made a mental note to give the floors a thorough inspection later and then set about my normal routine, commencing with the roof coverings.

The roof, upon closer inspection externally, had several cracked and broken tiles and one area seemed 'wavy' as though the timbers underneath had settled. I later discovered that most of the roof, when checked with a level, had been affected by slippage. But I digress, that was later on in the survey.

I worked my way around the property. The first real indication that something was wrong with the property came whilst I was inspecting the front wall of the house. The bay window had a distinct bulge in it around the DPC level. The walls when checked were found out of plumb. I knew that something was wrong but what was it? Once I had completed inside the roof, I asked if I could roll back the carpet. I never presume that I can take up carpet in case the edges are heavily tacked down. Mr Sly put up objections about having to move furniture but I then suggested that I would check in the corners of the rooms. Thus covered, I lifted a section of carpet in the dining room. Immediately I found large deep cracks in the floor that obviously not only went through the finish but into the concrete as well. The gaps ranged from ¼ in to ½ in in width. I knew immediately what I was looking at. It was a full blown **sulphate attack**. I have described sulphate attacks in detail later in the book, but briefly, this problem is caused when sulphates in the hardcore or the soil, travel to the concrete, carried in the ground water. As sulphates make concrete expand and break up, the problem is one that is not to be ignored. I rolled one section of carpet back and then went to look at the opposite side of the room and found a similar cracked surface. While this was going on the Slys remained in the living room. I considered the situation. I could tell Sly what I thought. It was obvious that he knew of the defects because in places the cracks had been partially filled up with plaster. Unless, of course, someone had done that filling without him knowing, which I found unlikely. I decided that Sly had hoped that I would miss the defects.

I considered some more. Should I confront Sly with the facts? In theory a surveyor could decline to tell a current owner about the defects found (reports being confidential) but I never have been able to keep completely silent, especially when a worried owner starts asking direct questions. Besides, once the haggling over price starts, the contents of the report will no doubt be bandied about. However, I needed an excuse to lift all the other carpets in the house. In the end, I decided to play the innocent and with a feigned worried look on my face, I went back in the lounge and told Sly about my 'discovery'.

Neither Sly nor his wife batted an eyelid. I asked if I could lift the carpet in the hall. Sly nodded. I guess by this stage that he was resigned to the fact that I wasn't going to be fobbed off with lies. I rolled back the hall carpet and found that the cracking ran the whole length of that room. I went back and told the Slys. It was then that Mrs Sly let the cat well and truly out of the bag by saying, 'Oh dear, is it in the hall now?'.

They had known, there was no doubt about it. It also accounted for why the bathroom floor had been replaced.

That night, I telephoned Mr Brown and gave him a verbal report and confirmed that I would send through my written report within the next couple of days.

Naturally, being concerned with buying a trouble free property, Mr Brown spent quite some time cross-examining me about the extent of the defect and, more important, the cost. I told him that, in my opinion, the only way to remedy the sulphate problem was to break up the existing floors and completely replace them. But, before embarking on such a drastic step, chemical analysis could be made on samples to prove that sulphates were the cause of the problem.

I later discovered that Mr Brown did not proceed with the purchase.

9 A client returns

(For full report see Appendix F: Snow Road)

Note
(See Fig. 9.1 for Mining Report which was supplied and is published with the permission of British Coal.)

Mr Brown is the type of client who makes it all seem worthwhile. As you will recall from the previous chapter, Mr Brown had been wanting to buy Fern Lea but decided against the purchase once he realized the extent of the sulphate attack on the floor. One never knows the client's reaction to a report unless they give you 'feedback'. In Mr Brown's case, the feedback was very positive. Although he was disappointed when the Fern Lea report turned out as it did, he was relieved that he hadn't purchased what would have proved to be a headache.

Some months later, Mr Brown made another appointment to see me and asked me to look at Snow Road.

Once again, Snow Road was a bungalow. As well as inspecting 6 Snow Road, Mr Brown wanted me to advise him if the loft could, if he wished it, be converted into living accommodation. This first requirement was answered within two minutes of visiting the house. As it was not possible to stand up in the loft, a conversion was out of the question (not unless the whole roof was rebuilt).

The bungalow concerned was a detached bungalow, with conventional cavity walls and a tiled roof. Other than the fact that in isolated places the damp proof course was slightly less than the recommended 6 in (150 mm) above ground level and the gas boiler needed attention, I could find very little wrong with the bungalow. It was obvious that the owner, Mr Fit, had looked after the property very well. When reading the report, note that under 'Main walls of house', I recommended that Mr Brown ask for any guarantees concerning the cavity foam filling.

This is a reasonable request and fairly important. In the early days of cavity filling, the 'cowboys' got in on the act and there were cases where cavity walls were made defective and let in water because of voids in the fill. In other cases, foam fill had been known to release poisonous gases.

Note also (see 'Building Control and Coal Board'), I was able to speak to

British Coal Corporation
Headquarters Technical Department
Mining Reports Office
Ashby Road, Burton on Trent,
Staffordshire, DE15 0QD
Telephone: 0283-550606
Telex: 34171 (CBTD G)
DX 29281 BRETBY

BUILDING SERVICES (TECHNICAL)
THE SQUARE,
TIPHAM

This matter is being dealt with
by
Survey Dept. (Tel. 0283-550606 Extn.

Our Ref: SR000001- - -

Your Ref:

Date: 2nd November 19_ _

Dear Sir,

Coal Mining Report
6, Snow Road, Tipham.

I refer to your enquiry dated 19th October 19_ _ in connection with the above.

Past Underground Coal Mining

British Coal have no record of any coal workings which could have affected the property.

Present Underground Coal Mining

There are no workings presently taking place within influencing distance of the property.

Future Underground Coal Mining

Although coal exists unworked in this locality, the possibility of future working is considered unlikely.

British Coal reserves the right to alter and amend its working proposals at any time should it be deemed necessary to do so.

Shafts and Adits

The property is clear of disused mine shafts and adits shown on our records.

Surface Geology

The property is clear of such faults, breaklines or fissures recorded on plans held by British Coal that may affect the stability of the property.

Claims for Subsidence Damage

British Coal have no record of having received a claim for subsidence damage in respect of this property in the last five years.

Opencast Coal Mining

The property is not situated within or adjoining an area for which an application has been made by British Coal for authorisation under the Opencast Coal Act 1958 or for planning permission under the Town and Country Planning Act 1971 in respect of the extraction of coal by opencast methods.

The property is not situated within or adjoining an area from which coal has been extracted by British Coal or their licensees by opencast methods.

Fig. 9.1   British Coal Mining Report.

Coal Act 1938 and Coal Industry Act 1975

The property does not lie within an area in respect of which a notice has been published under Paragraph 6 of the 2nd schedule to the Coal Act 1938 or Section 2 of the Coal Industry Act 1975.

Working Facilities Orders

So far as our records reveal, this property does not fall within the area of a Court Order made under the provisions of the Mining (Working Facilities and Support) Acts 1923 and 1925 and the Mining Industries Act 1926.

Payments to Owners of Former Copyhold Land

The property does not lie within an area in respect of which a notice has been published under Section 3 of the Coal Industry Act 1975.

British Coal have not received in respect of the property any notice of retained interests under Section 3(3)(b) of the 1975 Act.

Yours faithfully

A.H. Sturgess
Chief Surveyor and Minerals Manager

Fig. 9.1   Continued

a Building Control Officer who knew the area. Note that I qualified the information and confirmed that it was for guidance only. Note also the comments concerning fill ground. (See Fig. 9.1 for a typical Mining Report.)

It is worth studying the Mining Report and observing the sort of information that British Coal have in their files. Their records, however, in the main do not tend to extend back past the date of Nationalization.

This report is, I believe, a good point to finish Part One of the book because it goes to prove that Building Surveys do not necessarily have to be negative. Mr Brown bought the house in Snow Road, happy in the knowledge that it was unlikely to be expensive to maintain or cause him problems in the future.

PART TWO: Techniques and Considerations

10 Limitation clauses

(For reports see Appendices A to F)

There are six reports contained within Appendices A–F. You will note that certain clauses appear in each one.

The ones detailed below are not exhaustive and you are recommended to study the reports.

Those with a cynical nature may conclude that these are 'get out' clauses. This is not the case. These clauses should be included because the client has to realize that there are limitations in any survey and common sense dictates that these be brought to the client's attention.

I normally include most of the limitation clauses within the first section of the report under a heading 'Instructions, limitations and general preamble'.

There are other clauses that are contained within the body of the report. My main reason for not including them all under the main heading is to avoid making the first section too long and daunting for the reader.

Also some clauses are more logically included within the relevant section. Within the general preamble concerning instructions etc., I usually include the following clauses or one very similar.

CONDITIONS OF ENGAGEMENT

I usually ensure that a copy of the agreed terms are bound into the report and reference is made to them in the first clause.

CONFIDENTIALITY

A clause along the following lines is always included.

This report shall be for the private and confidential use of the client for whom the report is undertaken and must not be reproduced in whole or in part or relied upon by a third party. We will not be responsible to third parties who obtain, by any means, a copy of this report and act upon information contained therein.

Without this clause being inserted anyone, paying or otherwise, could if he or she obtained a copy of the report, sue the surveyor. Not possible?

Case law doesn't support this argument. Remember that a professional can be sued if someone acts upon advice given, even if a charge is not made.

CONCEALED FAULTS

Clauses along the following lines are always included.

Whilst every care has been taken in completing our survey and report, the investigations have been non-destructive in nature and therefore we are unable to report upon matters which are concealed at the time of the inspection and the following assumptions have had to be made:

(a) *That high alumina cement (HAC) concrete or calcium chloride additive or other deleterious materials or techniques were not used when the original property or any subsequent extensions were built.*

(b) *That any wall ties that exist are not perished.*

(c) *That your solicitor's or legal adviser's search will prove that the property is not subject to any unusual covenants, Local Authority restrictions, or onerous restrictions imposed by others, encumbrances or outgoings and that good title can be shown.*

(d) *That future inspections of any hidden parts of the structure which have not been inspected during this survey will not reveal material defects, or cause the surveyor to alter materially the indicative costings shown at the end of this report. (If applicable.)*

We have not inspected woodwork or other parts of the structure which are not uncovered, unexposed, or were inaccessible. Neither have we removed insulation from between or laid over ceiling joists to inspect the hidden surfaces of the timber because such an undertaking involves extensive builder's work which does not form part of our conditions of engagement. We are therefore unable to report that the property is free from defect. However, we have done our best to draw conclusions about the construction from surface evidence visible at the time of our inspection.

As you will note, this group of clauses covers the parts of the structure which either have not or could not be inspected during the survey. An unreasonable set of clauses? Unless you happen to have X-ray vision or take the whole house to pieces, it would be impossible to check everywhere. Even though endoscopes are becoming more common, and even if you were fortunate enough to carry one, there would still be places where inspections could not be made. You will also note that reference is made to HAC concrete and other deleterious materials.

Likewise the standard clause:

'*It was not possible to fully examine the floors etc.*' covers the obvious. With beds, heavy wardrobes and adhered floor finishes, how can the surveyor examine everywhere?

APPROXIMATE COSTINGS

Another useful clause is where costings have been included in the report:

As promised, we have included indicative costings to cover likely remedial works. However, we would stress that these are only intended to be an approximate guide and not an exact cost. They are given so that you understand the likely implications of the purchase in approximate monetary terms. As the costings are based upon average building prices, assumed specifications and assumed areas, some of the estimates may be high and some may be low. It is essential therefore that competitive estimates are obtained from several local builders prior to instructing work to be carried out. Our costings should only be treated as a guide, current at the time of preparing this report with no allowance made for value added tax.

Remember, without a clause like this the client might attempt to say that he was misled and that repairs have cost him in excess of what he anticipated. Note the reference to VAT. It is easy to forget this government surcharge.

THE ROOF

Under the section on the roof it is sensible to have a clause thus:

Where practical the interior of the main pitched roof has been examined but the inspection of the roof covering has only been made from ground level etc.

THE WALLS

When dealing with the walls of the house I always include the following:

No examination has been made of the foundations because this would require excavation.

Unless you have dug trial holes to inspect the foundations, then the client should be advised accordingly.

Another thought, would the average proud property owner allow a complete stranger to dig up his garden? Not likely is it?

THE DAMP PROOF COURSE

The damp proof course is such an important part of a property and it cannot be ignored.

There was evidence/no evidence of a damp proof course etc.

THE PROPERTY

Under the section that describes the property, I tend to put a clause:

We have not included superfluous information etc.

No doubt there are some who will say that this is wrong, but I know from experience and comments made to me by clients that they don't want 'padding'. They want me to show the 'meat in the sandwich'. In any case, most estate agents' information sheets cover such things as room sizes (Admittedly, there has been adverse comment in the media about room sizes in Estate Agents' advertising brochures, but even Estate Agents are human and electronic tape measures are in my opinion somewhat inaccurate in rooms containing furniture. Perhaps that will improve in the near future?)

SERVICES

NB If any service has not been tested, it is advisable to note this in the report.

RATING/PLANNING/BUILDING CONTROL/MINING REPORTS

Note also how I deal with rating, planning, Building Control and the Mining reports. Tell the client the extent of your enquiries/lack of enquiries. He/she has a right to know.

SKETCHES ETC.

Another set of useful clauses concerns the use of sketches and the position of right and left. I tend to include clauses along these lines:

Whilst we have tried to amplify our descriptions when technical terms have been used in this report, we would also direct you to our standard cross-section through a typical house and sketches . . . bound into the rear of the report.

References to left and right are made facing the house, standing in . . . The . . . for example is on the right of the house.

11 Building construction old and new

Note
The 1990/92 amendments to the Building Regulations have increased thermal regulations. In particular insulation is now required under floor – as most surveys will be carried out on houses built prior to this date, the new requirements have not been indicated on any floor details. However, loft insulation has been indicated as 150 mm thickness of glassfibre.

Where references are made to the National House-Builders Council publications the initials NHBC have been used.

GENERALLY

These notes are intended to provide outline information for those who are not familiar with domestic construction. Naturally, during a building survey, it will not be possible to see what is hidden from view, but knowing how the building should be constructed is obviously the first step in determining *if* it is built correctly.

FOUNDATIONS

The foundations of the house are probably one of the most important parts of the property because without an adequate base the property will quickly become unstable and collapse. Whilst it is usually impossible to inspect foundations during a survey, one has to know what should be there. It is also worth remembering that Building Control Departments sometimes have records of plans submitted. Subject to records being complete, one might be able to find records of what was constructed should there be any doubt regarding a building.

There are several types of foundation for new domestic construction which are:

(a) traditional strip
(b) deep strip

(c) tied footing or raft
(d) piled (very unlikely to be used).

Traditional strip

A typical traditional strip footing is indicated in Figs 11.1 and 11.2. Houses constructed under modern regulations would be built as per Table T2 of Approved Document 'A' (page 32) of the current Building Regulations. Bear in mind that older properties are probably not built to the new standards.

Deep strip

Similar to traditional, but deeper and probably 'trench fill', as Fig. 11.3 – used near trees. It is recommended that you refer to NHBC practice note 3 if there are trees nearer than 7.5 m to the foundations of the property.

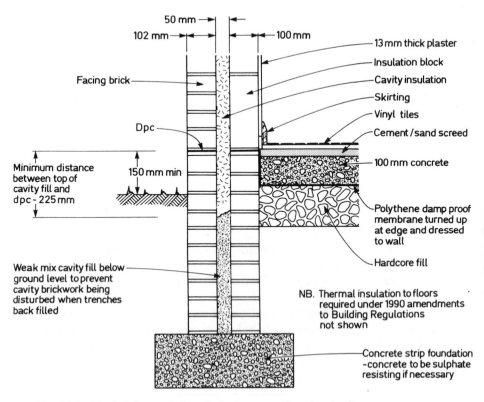

Fig. 11.1   Typical domestic foundation/concrete floor (modern).

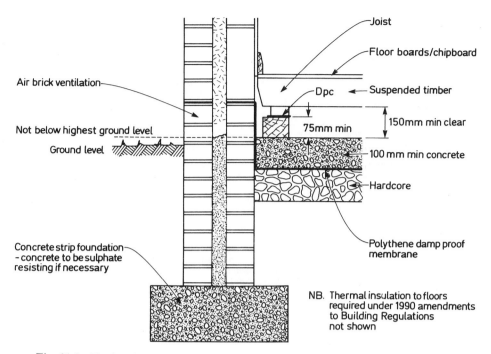

Fig. 11.2 Typical foundation/suspended timber floor.

Joist

Floor boards/chipboard

Air brick ventilation

Dpc

Suspended timber

Not below highest ground level

150mm min clear

75mm min

Ground level

100 mm min concrete

Hardcore

Concrete strip foundation - concrete to be sulphate resisting if necessary

Polythene damp proof membrane

NB. Thermal insulation to floors required under 1990 amendments to Building Regulations not shown

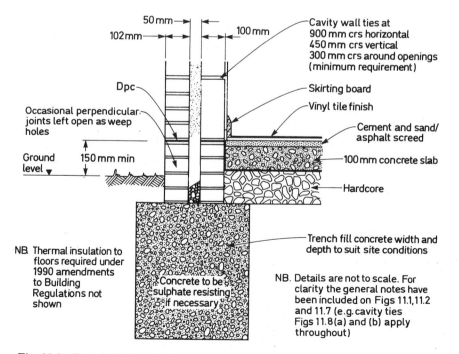

Fig. 11.3 Trench fill foundation.

50 mm

102 mm

100 mm

Cavity wall ties at 900 mm crs horizontal 450 mm crs vertical 300 mm crs around openings (minimum requirement)

Dpc

Skirting board

Vinyl tile finish

Occasional perpendicular joints left open as weep holes

Cement and sand/ asphalt screed

Ground level

150 mm min

100 mm concrete slab

Hardcore

Concrete to be sulphate resisting if necessary

Trench fill concrete width and depth to suit site conditions

NB. Thermal insulation to floors required under 1990 amendments to Building Regulations not shown

NB. Details are not to scale. For clarity the general notes have been included on Figs 11.1,11.2 and 11.7 (e.g. cavity ties Figs 11.8(a) and (b) apply throughout)

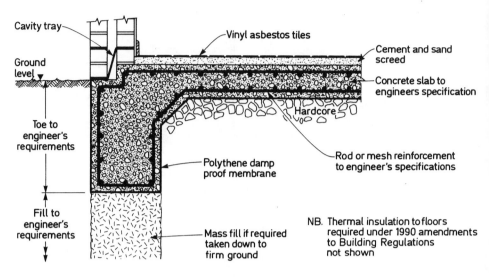

Fig. 11.4 Typical raft/tied footing (modern).

The tied footing or raft

A tied footing or raft foundation is a very useful form of foundation but does not come within the scope of the Building Regulations. The simple form of tied footing or raft which is met with in domestic construction usually comprises reinforced floor slab with a toe beam (Fig. 11.4). Because it is outside normal Building Control parameters, calculations are required by most Building Control departments before raft foundations can be approved. Once again, it is useful to know this fact because documents can be inspected in the Local Authority records.

The object of the true raft foundation is to spread the load of the structure above onto a very wide area of ground. Rafts are usually required where ground conditions are known to be bad, variable or in mining areas.

Older foundation types

In older properties built before modern Building Regulations came into effect the footing could well be built in very shallow trenches or off spread footings (see Figs 11.5 and 11.6). These types of construction are obviously less stable than modern strip footings. (*Shallow footings built on clay soil can suffer from frost heave.*)

WALLS (MODERN)

Most houses built under modern regulations will be of cavity construction. Cavity walls (Figs 11.1–11.3 and 11.7) below ground level usually comprise

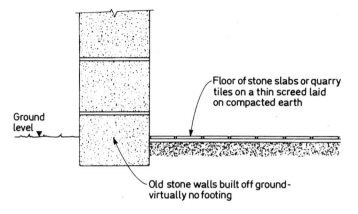

Floor of stone slabs or quarry
tiles on a thin screed laid
on compacted earth

Ground
level

Old stone walls built off ground–
virtually no footing

Fig. 11.5 Typical foundation in old cottage.

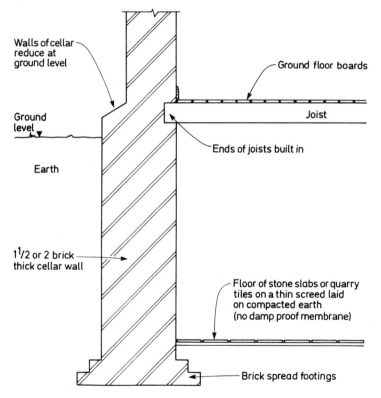

Walls of cellar
reduce at
ground level

Ground floor boards

Ground
level

Joist

Ends of joists built in

Earth

1¹/2 or 2 brick
thick cellar wall

Floor of stone slabs or quarry
tiles on a thin screed laid
on compacted earth
(no damp proof membrane)

Brick spread footings

Fig. 11.6 Typical old cellar detail.

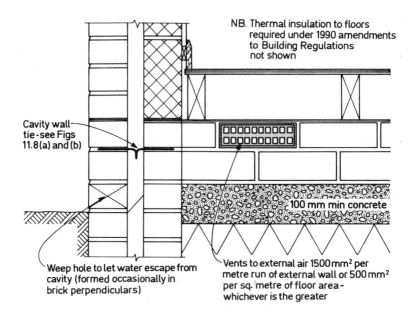

NB. Thermal insulation to floors required under 1990 amendments to Building Regulations not shown

Cavity wall tie-see Figs 11.8(a) and (b)

100 mm min concrete

Weep hole to let water escape from cavity (formed occasionally in brick perpendiculars)

Vents to external air 1500 mm² per metre run of external wall or 500 mm² per sq. metre of floor area - whichever is the greater

Fig. 11.7   Suspended floor showing ventilation (modern).

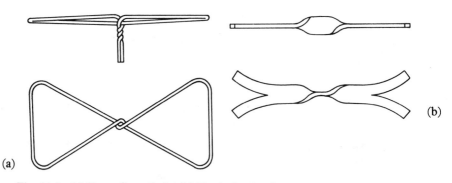

(a)

(b)

Fig. 11.8   (a) Butterfly wall tie. (b) Vertical twist tie.

two half brick skins with a weak concrete cavity fill between. Note that the cavity fill should be kept down 225 mm from the wall damp proof course. Above ground level, the two skins are usually a skin of facing brick, a skin of thermal block and a cavity between. Fears of an impending energy crisis have forced the government to pass regulations which are continuously upgrading thermal insulation in dwellings. The insulation factor is described as the 'U' value. If you want to know more about this topic, I would suggest that you refer to *The Building Regulations* or Pilkington Insulation's *Green Book Design Manual* (see Chapter 18, p. 167, for other useful texts).

Nowadays, it is becoming common to fill superstructure cavities with insulation (e.g. Drytherm). Under the 1990/92 amendments to the Building Regulations walls to habitable areas have to attain a 'U' value of 0.45 W/m²k (unless 'trade-offs' can be agreed – 0.60 W/m²k between a dwelling and a partially ventilated space such as a garage). The usual external skin of a cavity wall is in the facing brick. Stretcher bond on the external skin (Fig. 11.9) usually indicates the existence of a cavity wall.

WALLS (OLDER CONSTRUCTION)

Walls of older houses (Figs 11.5 and 11.6) are generally of solid construction. Solid walls are not as watertight or thermally resistant as cavity walls and this is why cavity construction is now the most common form of external wall used in domestic construction.

Solid walls can usually be noted because most cavity walls are built in stretcher bond (Fig. 11.9); solid walls are normally built in English, Flemish or English garden wall bonds (see Figs 4.7, 11.10 and 11.11).

In recent years it has become common practice to upgrade older 9 in walls by lining the inner face of the walls to increase the thermal insulation. However, if this process is not carried out correctly and vapour barriers and

The external appearance shown as all stretchers

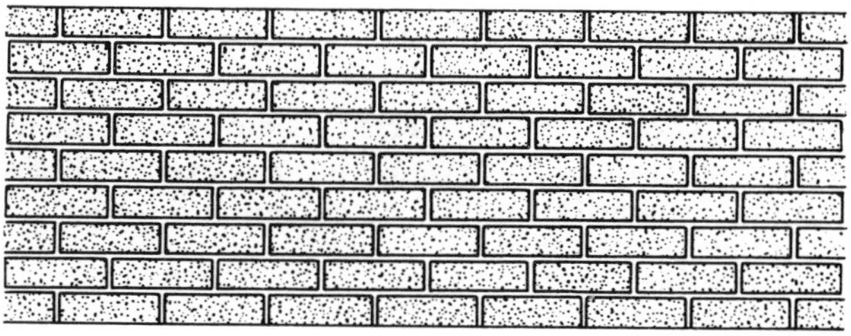

Fig. 11.9   Stretcher bond.

The external appearance shows alternate layers of headers and stretchers

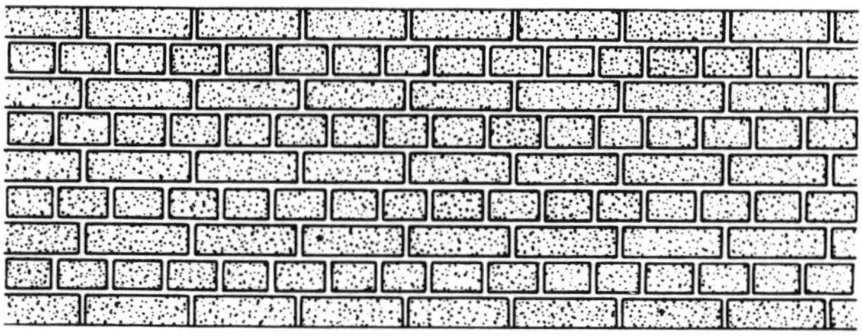

Fig. 11.10   English bond.

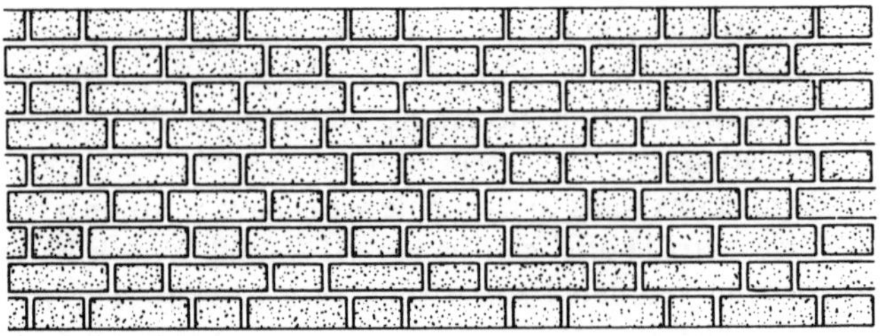

The external appearance
shows alternate headers
and stretchers in each
course

Fig. 11.11   Flemish bond.

the like installed, condensation can occur on the wall face behind the lining and create ideal conditions for woodrot. (See Chapters 2 and 12.)

OTHER TYPES OF WALL CONSTRUCTION

It would be impossible to describe every type of wall construction but here are a few:

(a) Solid wall rendered externally.
(b) Timber frame (beyond the scope of this book which deals with conventional construction, see Fig. 3.1).

Briefly in a modern timber frame house the main structural walls are timber-framed panels that are factory produced to a very strict design. The panels are highly insulated and provided with vapour barriers and breather paper. These panels are erected on site to form the walls. The floor and roof are then erected. If the house has any external brick walls, this is purely a feature; the brickwork has no structural function.

(c) No fines.

The no fines wall comprises pea gravel and cement mixed together but no sand added. During the construction period this no fines mix is then deposited between timber formers (formwork) and left to solidify. This type of construction was used extensively by Wimpey Construction some years ago to build council houses. The walls are fairly thick and provide insulation by virtue of the air trapped in the concrete. The no fines houses are rendered externally to provide water proofing.

LINTELS

The most usual type of lintels used on modern houses is metal 'Catnic' or similar type (see Fig. 11.12).

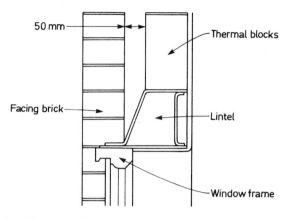

Fig. 11.12   Steel lintel detail.

NOTE :– All openings are to be continuous or provided by patent product which gives the same
area of ventilation. Details after Willan Building Products (Glidevale)

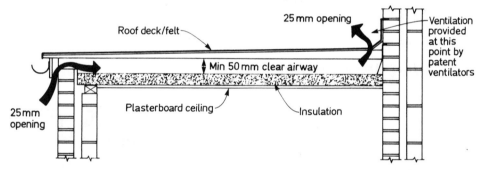

Fig. 11.13   Flat roof with abutments.

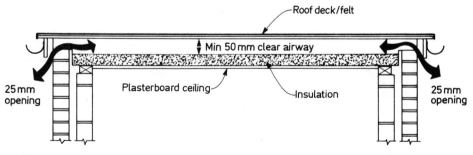

Fig. 11.14   Flat roof.

ROOFS GENERALLY

The 1990 amendments to the Building Regulations place great stress on providing ventilation within roof voids. This emphasis is because, with increasing insulation standards being demanded to conserve energy, a new problem has been created, namely condensation in the roof voids or on cold surfaces such as the roof boarding. It is worth bearing this in mind when conducting a survey. Figures 11.13–11.18 indicate the new ventilation standards. (The details are adaptations from catalogues kindly provided by Glidevale).

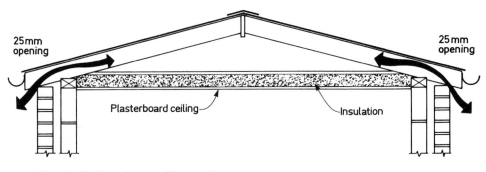

Fig. 11.15 Duopitch roof below 15 degrees.

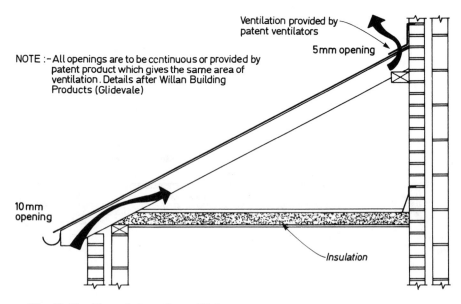

Fig. 11.16 Monopitch roof over 15 degrees.

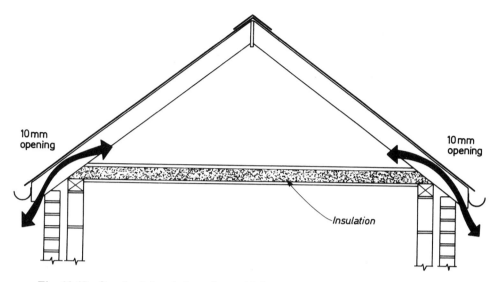

Fig. 11.17   Standard duopitch roof over 15 degrees.

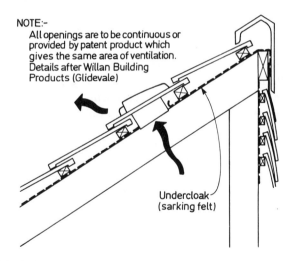

NOTE:-
All openings are to be continuous or
provided by patent product which
gives the same area of ventilation.
Details after Willan Building
Products (Glidevale)

Fig. 11.18   Typical modern ventilation for monopitch situations.

FLAT ROOF (MODERN)

A typical flat roof details are shown in Figs 11.19 and 11.20. Note the ventilation details.

Flat roofs are not (or should not be) totally flat. There should be a slight fall so that water drains off and does not pool.

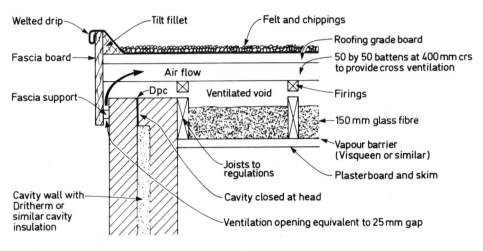

Fig. 11.19   Ventilated cold roof with tilt fillet fascia detail.

The roof details indicated are 'cold roof' type which is the cheapest type of flat roof. The roof is covered usually with three layers of bituminous felt (built up felt) which is covered by a 13 mm layer of white limestone chippings laid shoulder to shoulder in bitumen. The purpose of the chippings is to reflect the heat of the sun. Without them the roof will perish. The other alternative is to use solar reflective paint but as this is not accepted by most Building Control departments as having any fire rating, solar reflective paint finish can't be used near to boundaries with other properties.

Also note that a minimum of three layers of felt must be used. Where a prefelted chipboard is used, the prefelting must *not* be counted as one of the layers. With a cold roof, it is essential that the voids above the insulation be vented to the external air. Eaves ventilation should be provided as per the Building Regulations and provide 25 mm clear air space or its equivalent. There are now several manufacturers producing patent ventilators which satisfy requirements of the Building Regulations. Condensation will form on the roof deck and will cause staining on the ceiling below and eventually rot the roof joists and roof deck if adequate ventilation is not provided.

PITCHED ROOF (MODERN)

Refer to Figs 11.21–11.27. A pitched roof comprises a covering of concrete or clay tiles or slates. These must be laid to a suitable pitch (i.e. not less than the manufacturer's recommended slope or pitch). The roof tiles and slates are fixed to battens over sarking felt. The whole structure is supported by rafters the size of which can be found in the schedules in the

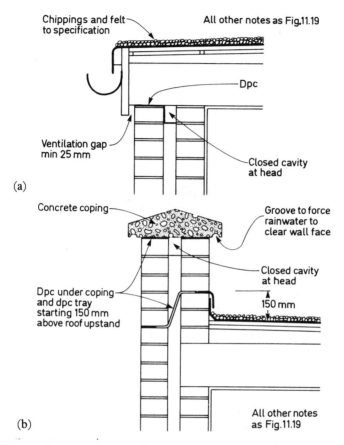

Fig. 11.20 (a) Ventilated cold roof details with gutter. (b) Ventilated cold roof details with parapet.

Building Regulations. For roofs of any reasonable span, the ceiling joists act as ties and prevent the roof spreading. The size of ceiling joists can also be obtained from the Building Regulations. The detail in Fig. 11.22 shows a roof of small span but on larger spans it is necessary to insert purlins (Fig. 11.21) and, once again, details of sizes can be obtained from the Building Regulations. As with flat roofs, under the 1990 amendments to the Building Regulations, new dwellings must have 150 mm of glassfibre quilt incorporated between the ceiling joists (laid over the plasterboard soffit (ceiling). (Alternatives can be provided if they give the equivalent 'U' value.)

Under the modern Building Regulations, the roof members also have to be strapped down to prevent any chance of a high wind damaging the structure.

On some large estates, builders use trussed rafters. These are computer

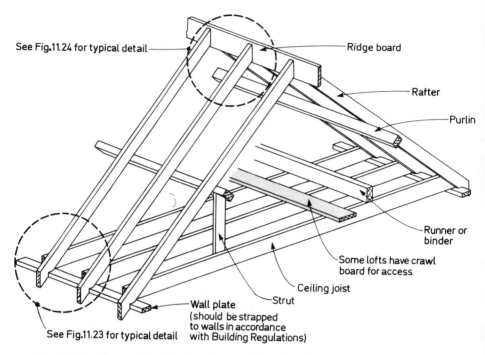

See Fig.11.24 for typical detail

Ridge board

Rafter

Purlin

Runner or binder

Some lofts have crawl board for access

Ceiling joist

Strut

Wall plate
(should be strapped
to walls in accordance
with Building Regulations)

See Fig.11.23 for typical detail

Fig. 11.21   Names of roof timbers in a traditional roof.

designed and factory made. However, to comply with modern regulations adequate bracing has to be incorporated into the structure. The National House-Builders Council (NHBC) have schedules of recommendations for bracing new roofs. It is a common fault for builders to skimp on this bracing in trussed rafter roofs.

PITCHED ROOF (OLDER CONSTRUCTION)

(See Fig. 11.21)

The slates or tiles did not have underfelt below the covering. Instead a cement and sand or haired lime plaster was smeared underneath the slates to waterproof the construction. It is very common to find that this render has fallen away.

SOLID FLOOR (MODERN)

The most common type of floor construction being used in modern houses in England and Wales is the solid floor. This comprises a hardcore bed (no

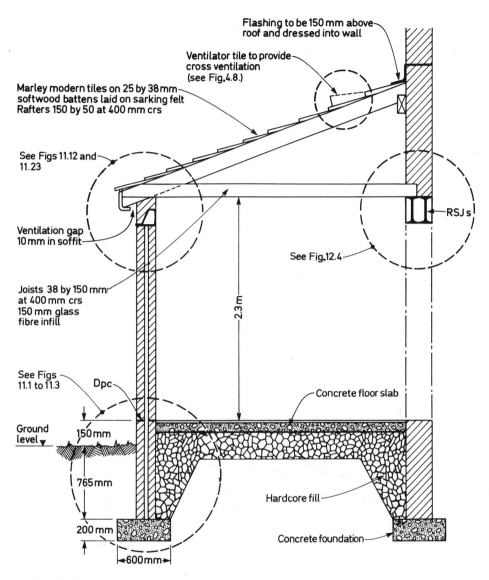

Flashing to be 150 mm above roof and dressed into wall

Ventilator tile to provide cross ventilation (see Fig.4.8.)

Marley modern tiles on 25 by 38 mm softwood battens laid on sarking felt
Rafters 150 by 50 at 400 mm crs

See Figs 11.12 and 11.23

Ventilation gap 10 mm in soffit

RSJs

See Fig.12.4

Joists 38 by 150 mm at 400 mm crs 150 mm glass fibre infill

2.3 m

See Figs 11.1 to 11.3

Dpc

Concrete floor slab

Ground level

150 mm

765 mm

Hardcore fill

200 mm

Concrete foundation

600 mm

Fig. 11.22   Typical section through lean-to extension showing pitched roof over 15 degrees.

thickness is specified but a minimum of 150 mm (6 in) is practical). For those who do not know, hardcore is crushed brick or crushed stone levelled and compacted as a bed for concrete.

It is *essential* that rubbish is not used as hardcore. The hardcore must not contain sulphates, bits of wood, plaster or other impurities which could

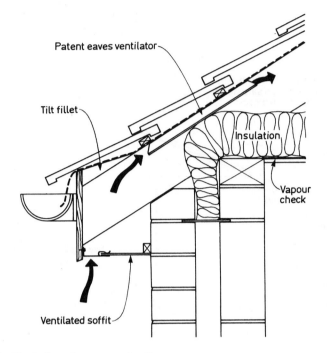

Fig. 11.23 Typical modern eaves detail.

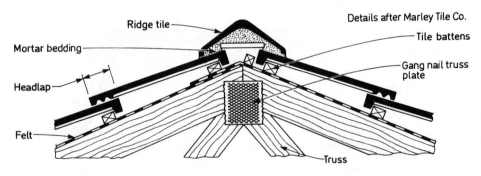

Fig. 11.24 Typical ridge.

deteriorate or attack the concrete bed which is laid on it. Old crushed bricks used as hardcore can also cause problems as the fill can introduce dry rot spores into a new house. The hardcore is blinded with fine material, usually sand, which prevents the 1200 gauge visqueen from being punctured (Figs 11.1–11.3). *Ash must not be used for blinding as it will contain sulphates which can attack the floor slab above.* A 100 mm mix ST2 concrete bed is

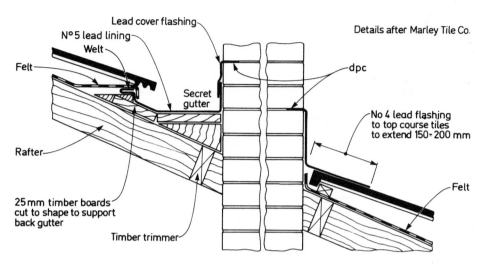

Fig. 11.25   Vertical abutment with secret gutter.

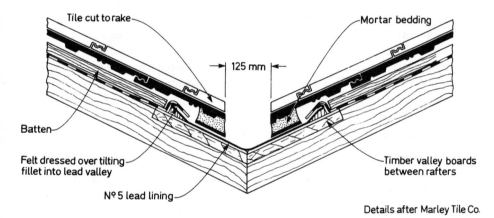

Fig. 11.26   Lead lined valley.

then laid on the visqueen and a 50 mm cement and sand screed laid on top of that. One commonly used alternative to a cement and sand screed is flooring grade asphalt. *The floor slab should not be lower than external ground level otherwise the house will flood in wet weather.* Where the hardcore beneath a solid floor exceeds 600 mm in thickness, the builder should have complied with the *NHBC Standards*, otherwise the concrete floor slab will settle.

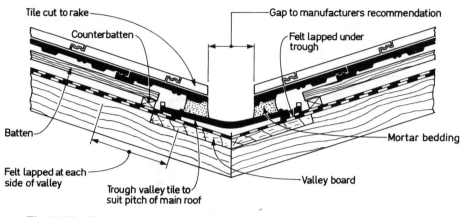

Fig. 11.27   Valley with special tile.

SOLID FLOOR (OLDER CONSTRUCTION)

Older properties quite often had red quarry tiles in the kitchen and scullery areas. These floors comprised 15 mm clay (quarry tiles) laid on lime/sand mortar direct on the earth. These floors are always suspect and should be replaced (Figs 11.5 and 11.6). The main dangers of this type of floor is that lino or vinyl sheeting if it is laid over the tiles can force moisture to build up under the covering. This process is known as sweating. The excess moisture can encourage dry rot and wet rot outbreaks.

SUSPENDED TIMBER FLOOR (MODERN)

A timber floor still requires a hardcore bed and a 100 mm concrete slab (Figs 11.2 and 11.7), which is why it is more expensive to construct in normal conditions. (If a large amount of fill is required because of site levels, then it could be cheaper to put in a timber floor.) Some people prefer timber floors as they are not as hard on the legs if one is standing up a great deal.

The sleeper walls are constructed off the concrete slab and wall plates and joists built off them. Note the DPCs and essential dimensions. *Once again, the concrete subfloor surface should be above ground level.*

NB The voids under timber floors have to be provided with ventilation to the outside air. This is done by installing air bricks. These should not be built in too low, otherwise water will flood into the house in wet weather. The purpose of airbricks beneath a timber floor is to prevent dry rot spores or other types of rot germinating on the wood of the floors.

Airbricks should provide 1500 mm² of free air space per metre run of external wall or 500 mm²/m² of floor area, whichever is greater.

Tanalizing timber (or similar treatment) will also help to stop woodrot outbreaks, a process whereby the timber is pressure impregnated with preservative.

PATENT SUSPENDED FLOORS

Some estates of modern houses are constructed using concrete beams infilled with blocks. One of the most common is the Bison Housefloor system. The main advantage to the builder of this system is speed of construction. Without prior knowledge or plans, it probably would not be possible to tell that the floor was not of standard concrete construction.

12 How buildings fail

Refer to Fig. 2 for pictorial illustration.

There are many ways that a building can fail. I have listed a few below:

(a) Obvious design faults
(b) Erosion and ageing
(c) Subsidence/impact
(d) Storm damage
(e) Water penetration/damage
(f) Biological agencies.

In the main, the building under attack is affected by one or more of the items listed above. Then water enters (e) and the biological agencies (f) then move in to destroy the timber.

OBVIOUS DESIGN FAULTS

Generally

Modern building practice in the UK has evolved over centuries. The modern 1985 Building Regulations (and 1990/92 amendments) and good trade practices are to a large part still based upon this continuous process.

If architects, surveyors and builders ignore the experience of generations or push manufactures' tolerances to the limits or fail to comply with Building Regulations, then it is almost inevitable that the faults that they put into buildings will destroy them, unless remedied.

For instance, I know of a whole estate of houses where the roof tiles were laid to the absolute minimum pitch recommended by the tile manufacturer but the roofers caused a defect by putting an extra large tilt fillet beneath the eaves tiles (see Fig. 11.23). All the houses on the estate have suffered from leaking roofs and woodrot because of this fault.

The list of all possible design faults would undoubtedly be a very long one and I do not intend to try to attempt to cover every conceivable option. If you refer to Figs 1, 2.1, 4.4, 12.2, 12.3 and 12.5 some design faults

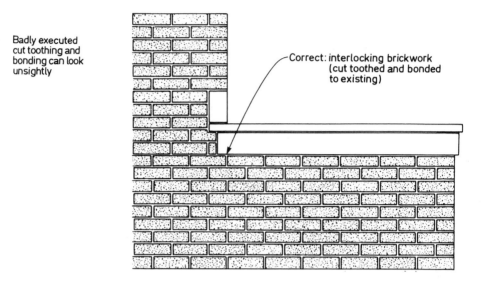

Badly executed
cut toothing and
bonding can look
unsightly

Correct: interlocking brickwork
(cut toothed and bonded
to existing)

Fig. 12.1   Correct vertical abutment.

Vertical abutments.   Where new walls abut old walls the brickwork should be cut, toothed
and bonded to the existing. In other words, pockets cut in the old
brickwork and the new bricks let in

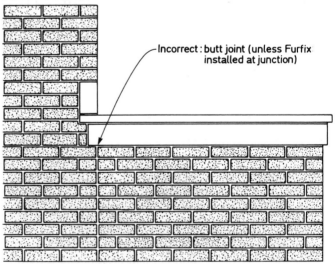

Incorrect: butt joint (unless Furfix
installed at junction)

Fig. 12.2   Incorrect vertical abutment.

become apparent. Note in particular Figs 12.1 and 12.2: extensions should
be correctly bonded to the existing building and not merely butted up.

I have nominated three faults for consideration:

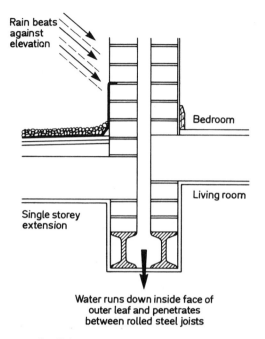

Rain beats against elevation

Bedroom

Living room

Single storey extension

Water runs down inside face of
outer leaf and penetrates
between rolled steel joists

Fig. 12.3   Incorrect detail for steel beams.

(a) No DPC/bridged DPC
(b) Cavity tie corrosion
(c) Incompatible materials.

No DPC/bridged DPC

Why is a damp proof course (DPC) needed? The answer is that without one the walls will draw up water by capillary attraction.

If you have ever seen the effect of rising damp, you will understand the reason that DPCs are essential. Without a damp proof course, the damp will eventually creep up the wall plaster to about a height of 1 m (3 ft) above floor level and create a stain in the decoration. This stain mark is of course something to look for internally whilst you are carrying out a survey. If the house has a 'tide mark' on the ground floor, it is a good indication that all is not well.

The Ancient Romans used a primitive type of DPC in their more prestigious homes. However, as with most of the ideas brought to this country all those centuries ago, the dark ages extinguished the light of that long defunct empire. The knowledge was lost for centuries and was re-introduced into the UK only in very recent times.

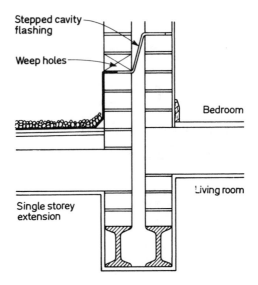

Fig. 12.4 Correct detail for steel beams.

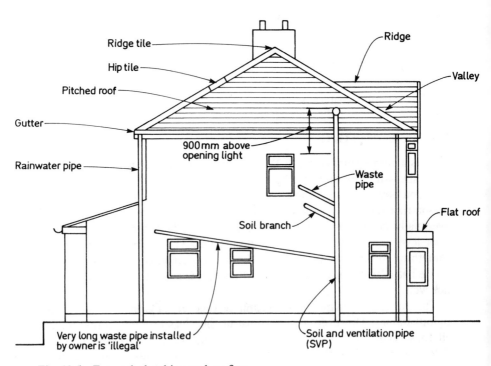

Fig. 12.5 External plumbing and roofing.

Most houses built in the UK since the 1920s have cavity walls and DPCs in the walls. Those built prior to this date are automatically suspect. (Those with DPCs should not be automatically accepted. Materials age and deteriorate. The DPCs in older properties do become less and less effective with the passage of time.)

In new construction, the modern DPC is usually composed of one of the following:

- A layer of patent damp proofing material (e.g. Hyload DPC).
- Stout polythene (used mainly in floors).

In older properties it is not uncommon to see:

- Slate (easily noted because of the wide mortar joints).
- Three layers of Engineering bricks.

In older properties that have been modernized the replacement DPC is usually:

- Injection or gravity-fed DPC (look for a line of holes or caps 150 mm above ground level, this is a tell-tale sign of injection work).
- Electro DPC (now falling into disfavour, this type of DPC uses an electric current to prevent damp rising but needs maintenance otherwise it can become defective).

A modern house usually has a patent DPC. Under the modern Building Regulations, the DPC should be situated a minimum of 150 mm (6 in) above external ground level; any lower than 150 mm and rainwater could splash up above the protective layer.

When inspecting the DPC, you should also check that the 150 mm minimum is maintained and that no one has built flowerbeds against the wall of the house so covering up the DPC. Another thing to watch for is that the path levels have not been raised. I have seen cases where someone has laid a new path on top of the old one. Double flags can usually be discovered by looking down into the rainwater gullies. Can a double layer be seen? Is the grating set down into a hollow?

In older houses the DPC may have been installed but is now no longer effective. A test with a moisture meter can help to indicate if a DPC is effective. Moss on the brickwork above the DPC is also a good indication of a defective DPC.

Cavity wall tie corrosion

As previously indicated, the modern form of external wall construction for houses is the cavity wall which comprises (usually) an external leaf of facing brick, a cavity and an internal skin of either brick or thermal blockwork.

The two 'skins' are held together with wall ties. There are two types of tie

Possible signs of wall tie corrosion may be observed as follows:

1. Gaps in the mortar joints visible with the naked eye horizontally and vertically

2. Wide mortar joints vertically 18 in (450 mm) apart

3. Movement of brickwork at door and window reveals

4. Bulging of brickwork and or rendering

5. Evidence of roof structure lifting

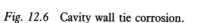

Mainly horizontal cracks – some vertical

Extra wide joint

18 in (450 mm) approx

Extra wide joint

Fig. 12.6 Cavity wall tie corrosion.

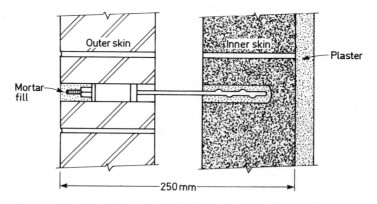

Outer skin

Inner skin

Plaster

Mortar fill

250 mm

Fig. 12.7 Brick to low density blockwork inner leaf – replacement tie.

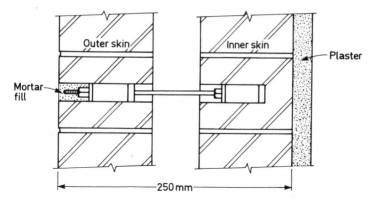

Fig. 12.8 Brick to brick inner leaf – replacement tie.

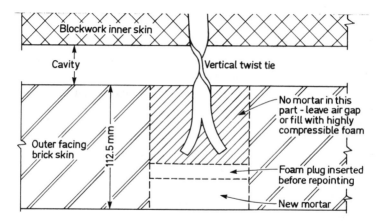

Fig. 12.9 Plan view of typical cavity wall showing ends of existing wall tie isolated.

in common use, the butterfly and the vertical twist type (see Fig. 11.8(a) and (b)). These ties were at one time made of galvanized steel. The zinc layer on the cheaper ties was very thin and over a period of years the zinc tends to degrade leaving the steel to rust. The loss of the zinc and subsequent rusting can be accelerated if the wall ties were built into walls that had been built in 'aggressive' mortar. Once the metal in the ties starts to turn into rust, an expansion process starts. The total vertical expansion in a two storey building can be as much as 50 mm. This sort of damage does not normally occur when light 'butterfly' ties have been used. The visible effects are horizontal cracking in the mortar joints on the outer leaf of brickwork coinciding with the level of the ties which are usually four courses apart. The cracks are usually several millimetres wide and mainly on the horizontal joints (see Fig. 12.6).

Without signs of corrosion being obvious, the only way to detect it is by using an endoscope or by removing one or two bricks and inspecting the ties. However, one has to be careful. Both techniques are destructive (i.e. one has either to drill a hole for the endoscope or cut a brick out). Unless the owner agrees, you cannot carry out such an inspection.

I think it is worth mentioning *Building Research Establishment Digest 329* at this point. It is obviously very expensive taking down the whole outer skin of a house to rectify cavity wall ties. This digest describes a way of installing wall ties in existing walls using the isolation technique. Using this method, the old tie is left in the wall but the end of the tie in the external skin is isolated by chiselling out a plug of mortar around the end of the old tie. Foam plugs are put in and the joint repointed. The foam plugs prevent any further expansion damage. New purpose made stainless steel ties are then inserted in new locations in the wall (see Figs 12.7–12.9). However, this technique is not recommended for walls with thin mortar joints as there will not be sufficient room for the technique to work. (A normal joint is 10 mm.) Where the joints are too thin, then the only method of rectifying the damage is by cutting the old ties out by removing bricks or occasionally by rebuilding the whole wall.

When inspecting a property that has supposedly been re-tied using this system look for small mortar 'dots' on the brickwork and that the old ties have been isolated. Make sure that beds are at least 10 mm wide as per BRS 329 recommendations.

As recent reports have indicated that most houses with ties have some corrosion problems and as there are an estimated 12 million houses in Britain with cavity walls, one of the most common problems has to be corroding wall ties. Already, more and more building societies are asking for checks to be carried out on wall ties when properties are sold. It is therefore not surprising that cavity wall tie replacement has become a real growth industry in the UK.

Incompatible materials

There are many instances of materials being used in the wrong place. The sulphate attack on the floors of Fern Lea is an example:

Sulphate attack on concrete floor
Sometimes builders have been known to use ash or other deleterious material below concrete floors. Eventually water will reach the ash and release any water soluble sulphates contained therein. Unless sulphate re-sisting concrete has been used in the floor slab, the concrete will be attacked.

The tell tale signs of a sulphate attack is indicated by a lifting of the floor which then causes doors to bind at the bottom edge. As the concrete is further attacked, there can be lifting and arching of the surface and cracks form in the upper surface. The external walls can also be moved outwards near to the damp proof course level. If one finds the outer wall overhanging the DPC, be suspicious.

EROSION AND AGEING

Human kind have never really accepted that they are controlled to a large extent by the same rules that govern the whole universe.

The truth is that Mother Nature, red in tooth and claw, takes no prisoners. She doesn't make any distinction between the works of human kind and any other part of the landscape.

Hills are born and then eroded. Trees grow, live and then die. Buildings are no different. Unless they are constantly repaired and maintained, then the materials that compose the building will decay. Even as the first brick is being laid on a new building, the elements, the wind, rain and frosts, are there eroding the exposed surfaces.

This phenomenon is really brought home when a large building such as a cathedral is being built. It sometimes takes decades to complete the work. It has been known for parts of the building to need repairing even before the last brick or block of stone is laid.

SUBSIDENCE/IMPACT

This section is fairly self-explanatory. If the foundations give way because of a landslide or if a lorry runs into a property, it will be damaged. However, under this heading also comes mining and tree damage.

Mining and salt extraction

Mining areas should be treated with extreme caution. I recently visited one area where mining had been pursued for many years. It was obvious that most houses in the area had been affected in some way. The number of houses that had had to have gables rebuilt were incredible. Towns like Northwich should also be treated with caution. Brine pumping is a major industry in this town and subsidence is extremely common.

It is worth remembering that it is possible to obtain Mining Reports from British Coal which will advise whether mining has occurred in the area where the property has been built and if it is likely in the future (see Fig. 9.1).

Trees

Trees growing close to houses must always be treated with suspicion. The National House-Builders Council has recognized this problem and has

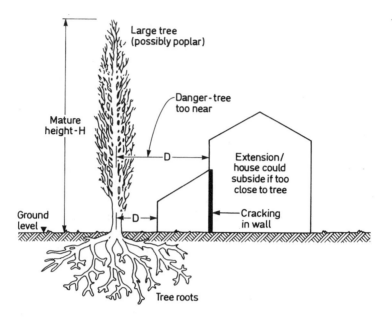

Fig. 12.10 Damage caused by trees.

published their recommendations in *NHBC Standards – Building Near Trees* (formerly Practice Note 3). When houses are built on areas of shrinkable clay and a tree is too close, it is likely in very dry weather that it will suck the moisture out of the clay beneath the foundations which then move downwards (Fig. 12.10). Obviously this downward movement stresses the substructure and cracking is likely to be visible as the ground falls away. The cracking is predominantly diagonal and follows the horizontal and vertical mortar joints in the brickwork. The cracks are widest at the top and narrowest at the bottom. The cracks also tend to open and close during the course of the year depending upon rainfall. (This opening and closing is obviously not instantaneous, but measurements taken over a period of time can detect the movement.) Doors and window frame distortion is usually noticeable where subsidence has taken place.

Returning to *NHBC Standards*, recommendations are made concerning distances of trees from houses. Unless one is certain that the house has non-traditional foundations (i.e. deep strip or a raft), I would recommend that the distance and height of trees in gardens be checked.

The NHBC recommend that the mature height of the tree be assessed and compared with the distance to the house. Working on the principle that most foundations are only 1 m deep, the tree should not be closer than its likely mature height (e.g. if the tree is likely to be 30 ft (10 m) high, then it should be 30 ft (10 m) away from the house) if there is any likelihood that the house is built on shrinkable clay.

If the house has deep footings, then closer distances may be acceptable. I would recommend that you obtain a copy of the *NHBC Standards* and peruse the details.

STORM DAMAGE

Once again, this is self-explanatory. If storm damage is not repaired promptly, then damp and biological agencies will inevitably enter the building (see below).

WATER PENETRATION/DAMAGE

Probably the most important requirement for any building in the UK is that it is weatherproof and damp resistant.

The following are the main causes of dampness in buildings:

(a) There is tile/slate slippage or other defects in the roof coverings.
(b) If it is a felt roof the covering material fails or the design is defective.
(c) The flashings are missing (e.g. at abutments or around chimneys).
(d) The flaunching to the chimney is defective.
(e) The gutters and downpipes are either broken or discharging water into the building (check especially the backs of the old cast iron pipes).
(f) Valley and/or secret gutters are leaking.
(g) In cavity wall construction cavities in walls are defective and water is entering the building (i.e. it is coming along the wall ties) because of mortar droppings.
(h) In solid walls the render/pointing is defective and water is entering the building.
(i) There are plumbing leaks in the kitchen/bathroom.
(j) Condensation is forming in pitched roof voids caused by lack of ventilation.
(k) Condensation in flat roof voids caused by lack of ventilation.
(l) The damp proof course either does not exist or is defective.
(m) The damp proof course has been bridged by flags or earth.
(n) The air vents under floors are blocked and moisture is building up in the trapped air.

One has to be on the lookout for signs of water damage/dampness while inspecting the property. Watch out for dark stains on wallpaper, efflorescence (i.e. white powdery patches) on plaster, mould growths or moss. The reason for excessive dampness being a problem is that plaster can crumble and overwet bricks can spall (i.e. the bricks break up as a result of frost action).

In timber, excessive damp can lead to outbreaks of wet and/or dry rot. Once dampness is noted, it must not be assumed that the problem is in the immediate vicinity. Don't forget, water runs and percolates and it does not show up exactly adjacent to an external fault. You have to apply some logic when trying to trace a fault. Dampness on an upper ceiling would indicate a roof leak whereas a damp patch on a wall could suggest defective brickwork. Signs of damp on a ground floor ceiling well away from an external wall could be caused by a leaking pipe. If dampness is noted on the ground floor near to the skirting, it is more than likely to be rising damp.

Once water gets in by whatever means, then the conditions are ripe for the biological agencies to get to work (see below).

BIOLOGICAL AGENCIES

See Plates A to H for visual identification, pp. 116-117

This section can be further subdivided as follows:

(a) Dry rot
(b) Wet Rot
(c) Woodworm.

Dry rot/wet rot and woodworm – generally

Although concrete, steel and plastic are being used more and more in modern buildings, timber remains one of the most important materials used in houses, the reason being that it is relatively cheap, versatile and easily worked. However, Nature draws no distinction between a piece of dead wood in a forest or in a house. When conditions are right, these biological agencies will germinate in houses and destroy the timber. The terminology used in describing different types of wood decay is somewhat ambigious. Woodrot is often spoken of as wet rot and dry rot. Since both types of rot can only attack wood when it is damp, the division is misleading. True dry rot is *Serpula lacrymans* (formerly *Merulius lacrymans*), it eventually leaves the wood in a dry, friable condition but the fungus needs damp, ill-ventilated places in order to grow and, unless it is allowed to establish in one area, it cannot attack dry wood.

The general term 'wet rot' covers several other types of wood destroying fungus rot.

For the average surveyor, the most important point is to be able to distinguish between wet and dry rot. The reason is that wet rot is relatively easy to treat. Dry rot should really only be treated by a specialist and is therefore usually very expensive to kill.

Dry rot (Serpula lacrymans)
See Plates A to E, pp. 116-117

Dry rot is a wood destroying fungus which lives by eating the cellulose in the wood. Dry rot is the most serious fungus attack that a building in the

UK can sustain. True dry rot normally requires a moisture in the wood of over 20% for the fungus spores to develop. Once germinated, hyphae (thin greyish strands of the fungus) grow from the spore in order to spread over and through the wood. Eventually, the hyphae will become so dense as to form a sheet of growing fungus known as a mycelium. The colour of the mycelium tends to vary in colour from dirty grey to pure white in wet conditions.

Once the fungus, like any plant, has extracted enough nourishment to grow, it fruits by forming a sporophore which gives off millions of red dustlike spores (seeds).

Dry rot gets its name because the wood attacked becomes dry and brittle with a 'charred' appearance as the timber forms cuboidal fractures. When destroyed, the remnants of the timber crumble easily in the hand. Dry rot fungus spores are so fine that they are in the atmosphere everywhere, just waiting for the right conditions in which to germinate. The conditions favoured by dry rot are unventilated areas with timber moisture content above 20% and an air temperature of around 20°C. Unlike other fungus, dry rot will send its tendrils through non-edible material such as brickwork, plaster or sometimes even cracks in concrete in order to search out new supplies of wood to consume. The new wood doesn't even have to be damp because once established, dry rot can extract moisture from the surrounding air and then take it to the new wood in order to feed. As the fungus eats the cellulose in the wood, the wood loses strength, cracks and shrinks. Damp cellars, underfloor areas, kitchen sinks and around sanitary fittings are the most common places to find outbreaks. The tell-tale signs of dry rot are fungal growths, obviously decayed timber, and discolouration or ir-regularities in the wood. The fungus itself can be recognized by the fact that the fruiting body of the fungus is a rust red colour and the threadlike strands are grey white. Dry rot also smells musty or very 'mushroomy' if disturbed.

Eradicating dry rot can be very expensive because it is not just a question of merely cutting away affected timber. In order to completely avoid the risk of the infection spreading from the last noted outbreak at least 1 m of timber must be cut away and burned.

Any remaining timbers in the area and other areas must be treated. In particular, in order to kill strands remaining in brick walls a chemical barrier has to be formed by irrigating the walls. Plaster has to be hacked off and sometimes a blow torch applied to kill strands. (Heat treatment is not common now because of fire risks involved and also because it is not possible to know exactly when the killing temperature has been reached.)

Wet rot

This term covers several fungal types the most common are as follows:

Coniophora puteana (formerly *Conrophora cerabela*). *Coniophora puteana* is not as serious as dry rot. In order to germinate, moisture in the wood must be between 40 and 50%. When moisture is removed, the rot growth ceases. Unlike dry rot, wet rot mycelium does not really spread over non-wooden elements in search of new timber, and if it does, the distances travelled are much shorter. (I have seen an example of this fungus creeping out a distance of 2 ft to dine on a cardboard box left lying on a cellar floor.) The affected timber goes a dark brown colour and breaks up into small cuboidal splits or longitudal cracks. (See Plates F and G.)

Phellinius contiguus. This is a white rot eating both the cellulose and lignin of timber. Because of the poor quality of modern softwood timber, it has in recent years become the cause of decay in external joinery.

Woodworm. These wood boring beetles start life as an egg laid on a wood crevice. After hatching the grub eats its way into and through the timber and continues to do so until it pupates. When it turns into a beetle, it bores its way out leaving a characteristic flight hole.

Common furniture beetle (Anobium punctatum). This is the most common woodworm in the UK in rural areas it is estimated that 80% of houses over 40 years old have some woodworm. Woodworm prefer damp timber. It is therefore possible to find timbers with woodrot and woodworm in the same areas.

As suspended timber ground floors, bathroom floors and floors near WCs get wetter than other areas, these are also the most likely to be attacked.

The woodworm usually take three or more years in the wood burrowing stage. The beetles usually fly between May and September and fresh flight holes and frass indicate that infestations are still active. (See Plate H.)

13 Carrying out an inspection

These notes are only intended as a guide and have been written on the assumption that you have never carried out a survey before, or if you have, not many.

FOOD FOR THOUGHT

With the aid of the information contained within the previous chapters, you should now have an outline of how a survey is conducted and the defects that can be present. The purpose here is to bring previous chapters into perspective.

Before we start, let us consider some basic psychology. They say that moving house is one of the most traumatic experiences in life. I forget the score out of ten, but it is supposed to be very high on the scale. The person(s) selling the property are probably in a 'chain' and desperately don't want anything to break it.

In addition, they may have had several prospective purchasers express an interest in their present house and then either 'drop out' for no apparent reason or simply not have the courtesy to let them know that they are no longer interested. All the time there is the worry that the professionals who are dealing with their sale are 'trying to keep the clock running'. I know that in the main this sort of accusation is untrue, but to a worried householder it may feel like it.

Then there is just a possibility that they could be having difficulties obtaining a mortgage. Perhaps, when they first considered moving house, they were promised a mortgage, but now there seems to be a difficulty that their building society will not fully explain. Maybe an ultra-cautious valuer has suggested that there could be something wrong with the house that they would like to buy.

Then there is the reason for moving in the first place. Possibly the move is caused by a change of job, which may also create high stress factors.

Then there is simply the stress caused by people calling – or maybe the opposite. They have advertised their house and no one seems interested in buying it. They, in turn, may have spent long hours visiting estate agents followed by the grind of actually visiting other houses only to be disappointed by what they see.

Of course, there is also the legal system which has never been renowned for speed. All these factors create stress and worry.

And finally, the buyer says that he/she wants a Building Survey carrying out prior to contracts being signed.

I am sure that you will agree that it goes without saying that the surveyor's visit is not going to be greeted with joy. From the stories that I have related previously, the need for a survey is fairly obvious, but to the vendor whom you are visiting, this is irrelevant. An unknown person is going to come to their house and find fault with it!

I would suggest that a sensible surveyor ensures that he/she take these factors into account and acts with tact.

So let's start as we mean to go on. Don't be late for the appointment.

Okay, so the lady of the household answers the door when you call. She could be a housewife. However, in these days of high mortgages, she probably works and has had to take time off just so that she may let you in. It has cost her money just so that you can earn your living.

If a delay is unavoidable, a telephone call explaining the position will certainly be appreciated.

When you reach the door, be prepared to show some identification. I make a point of presenting one of my business cards as I introduce myself so that there can be no doubt as to my identity. It also creates a good impression.

Be polite and friendly from the outset. I always make the time to put the vendor at their ease by explaining what I need to do and why and approximately how long it will take. (You should have given some indication as to time when you confirmed the appointment but you are advised to reinforce what you have said.)

If possible, do give an indication of the likely sequence, such as, 'I'll be starting outside . . . that will take about half an hour and then I would like to go into the roof space . . . Is that alright?'

Once inside, I would suggest that a friendly approach is maintained. You should take the trouble to explain what you are going to do and not merely wander into every room without permission. I always make a point of letting the owner know when I am about to go upstairs. The upstairs of a house is usually the most private part.

Apologize in advance for any disruption that you may cause. If you are polite, then nine times out of ten the vendors will return that politeness. If you are rude, then they may respond in kind.

Most important, keep your inner thoughts to yourself and avoid making comments that would offend. After all, you are a guest in someone's house.

You are there purely to formulate a report on the fabric of the building and not on the social status of the occupants!

Also don't be scared of talking to the owners of the house or asking polite

Plate E
Dry rot spore dust is often deposited near fruit bodies.

Plate F
Coniophora puteana. Cellular fungus.

Plate G
Brown strands of *Coniophora*.

Plate H
Common furniture beetle damage
in ceiling joists.

Plate A
Dry rot strands in masonry.

Plate B
Dry rot mycelium under a floor.

Plate C
Dry rot attack of wood.

Plate D
Dry rot fruiting body.

questions. Don't forget, the people that you are visiting have probably been chewing their nails, wondering if you are going to find anything wrong. They will be far happier if you talk to them and treat them like human beings. I find that if you take the trouble to make them see that you have no 'axe to grind' and that your report will be factual, then they will accept your presence far more easily.

If you act in an unfriendly manner, simple information, such as the whereabouts of manholes and the like, will only be given grudgingly and a bad atmosphere will be created, which will not help you carry out the survey.

If you adopt a superior, condescending attitude, then they may take a strong dislike to you. If you are offered a cup of tea, accept. They are trying to break the ice.

If you take the trouble to prove that you are a fellow human being with a job to do, then they are more likely to be tolerant.

Obviously there are limits. Time is money. Also you have to be careful what you say, and your client's business must be kept private and confidential, but a smile goes a long way to making your task easier.

Respect their property. You need to use your moisture meter but try to take readings in discreet positions. Tell them why you are taking the tests and if necessary ask permission. The *I need to check out the skirting boards with this . . . You don't mind, do you?* approach is best.

If they do object to certain activities, I find that you can usually obtain consent by telling them that things may not appear satisfactory in the report if you don't. That usually persuades most people to give their blessing to what you propose to do.

They will be anxious, waiting with bated breath as you go around the house. If your tests prove that the house is in reasonable repair, then put their minds at rest but take care. You should not be too open, as your report is supposed to be confidential.

What I am trying to get across is that a good surveyor should, for the want of a better word, be able to *manipulate* the situation so that he/she obtains the co-operation of the owner.

Respect property. If you are testing a window cill, take the reading underneath if possible. Although you will need to probe test timbers, use your fingers first. Why damage paintwork when it is obvious that the wood is sound? If it feels soft to the touch, then test it with the probe.

As you will have noted, I go out of my way to find out as much as I can about properties by discussing matters with the owners.

Asking questions on the spot such as, 'Is there a guarantee for the DPC?' and stressing the benefit of there being one, usually brings results. The owner will often disappear into a cupboard to find the relevant piece of paper. The rest of the survey can continue while they search.

The concept of vendor co-operation can be taken one stage further. When writing to the property owners confirming the time of inspection, you can ask for details concerning possible extensions to the property, if Building Control and planning consents were obtained, the application numbers, if new DPCs have been installed, the name of the installer, guarantees, etc.

In my experience, most vendors will co-operate, especially if the initial letter intimates that they will be helping the surveyor to view the property in a better light if the information is to hand.

Even so, friendliness alone does not always work. Occasionally, you will come across an aggressive vendor.

I can recall one chap who tried to bully me around the property in question in ten minutes flat, muttering all the time, 'The other surveyor didn't take this long' and 'If Mr Smith thinks that he can get me to drop the price by you finding faults then he's mistaken'. In this case, I countered by asking questions about the other surveyor.

The vendor very quickly started to change tack. After all, the implication was that if the house was still on the market, then the other surveyor's report could not have been particularly positive.

I then followed up with *I'm only doing my job*, and that eventually won the day. With bad grace, he let me do what I needed to do.

Some owners fear that they will assist in their own downfall (i.e. that the surveyor will find something wrong with the house). For instance, I do usually ask, prior to visit, that the current owner releases carpet at edges to assist the survey. Some don't bother, saying that they were too firmly fixed. This approach can make the inspection of floor surfaces very difficult.

If I am in real doubt concerning the floor under a carpet, I make a point of expressing my misgivings to the owner. At that point, they then usually co-operate and find places for me to roll back the carpets. Obviously, they don't want a bad report.

As I have already said, it is prudent to advise how long you will be at the house. Time taken in carrying out a survey can vary considerably. Some surveyors work quickly, whilst others are plodders; some houses only have a few rooms, whilst others can be tortuous. In general terms, the older the property, the longer the survey. An old, largish terraced house, for instance, is far more likely to have defects than a small, brand new semi and the prudent surveyor bears this in mind.

In a largish terrace, I usually tell the owner that I will need 3 to 4 hours to carry out my inspection. I don't normally need so much time, but it is better to overestimate than underestimate.

As I have said before, I also explain what I will be doing (i.e. going into the loft, checking the DPC, going down into basements, etc.).

When making arrangements, take account of the time of year. It is no use starting a survey as the sun is setting on the horizon. Bad weather can be a friend. Don't become a fair-weather surveyor. A heavy downpour can be a blessing in disguise. Roof leaks and defective gutters can easily be noted when it is raining

Make sure that the survey is carried out in a formal way. Don't rely on your memory. Write down notes or tape record the details at the time of inspection.

(*Tape recorders can be a problem as the owner can hear what you are saying . . . I usually take myself off somewhere out of the way if I'm about to record a serious defect . . . It all comes back to not saying disagreeable things in front of people . . . I also take the precaution of taking a notepad with me in case it is too awkward to use the tape recorder.*)

As I have pointed out already, some vendors will follow you and keep a very close eye on your progress. Don't let them embarrass you!

With some, following is just curiosity, with others a deliberate attempt to put you off. Being on your own is best, but if you are going to have company use the time usefully by asking the vendor for information. (*In the nicest possible way, of course.*)

Most people who are just trying to be obstructive will quickly decide that they might give something away and will prefer to leave you on your own.

NOTES ON KITCHEN PLANNING

One of the problems of being a surveyor is that as soon as anyone finds out what you are, there is somtimes a tendency for them to drag you over to a wall and demand to know why they keep getting condensation on a particular surface, despite the fact that every builder in the area has so far tried to cure the defect.

Some years ago, I was introduced to a chap who related a tale of woe about the kitchen in the house that he'd just purchased.

I listened and took note. Apparently, the surveyor that he had engaged to survey his house had not mentioned in his report the poor state of the kitchen fittings. When the new lady of the house had come to stock her cupboards it became obvious that the doors were badly fitting. Naturally, they were not amused.

Now the kitchen is one of the main activity centres of any house. If the kitchen is inadequate, then it is a major source of annoyance.

Ever since the conversation I have made a point of ensuring that I open and close all the cupboard doors in kitchens, and make an overall appraisal of this important room.

Every kitchen should have space provided for a cooker, refrigerator, dishwasher, freezer, washing machine and tumble dryer.

When examining a kitchen look for the three main activity areas, namely:

(a) The food storage area (larder, fridge/freezer and/or store cupboard).
(b) The sink.
(c) The cooker.

In your mind's eye create a *work triangle* between these areas.

If you add together the sides of the triangle, in a well planned kitchen the length of the combined sides should be no longer than 6.6 m (21.6 in) and no shorter than 3.6 m (11.10 in).

If the dimensions are any bigger than these, it will result in unnecessary walking; if they are any smaller, the kitchen will be cramped and awkward.

It goes without saying that in a modern kitchen there should be plenty of good quality worktops and that these should link the working areas together.

It is rare, however, that anyone has an ideal kitchen because most builders construct the kitchen as an afterthought. Developers tempt their customers with large lounges, downstairs WC's, stained woodwork, open staircases and other fads as selling points and kitchens tend to come very much at the tail end.

Here are a few pointers as to what an ideal kitchen should be like.

(a) If possible it should have a double bowl sink, with one bowl capable of taking a waste disposal unit.
(b) Sinks should be:
   (i) located under windows;
   (ii) no further away from a soil stack or gully than 2300 mm;
   (iii) not sited in a corner of a room or up against a cupboard.
(c) There should be:
   (i) room left on the drainer side for some one to dry up.
   (ii) the drainer should be located on the side away from the cooker (with a single drainer unit) this will allow quick access to the sink with hot pans without having to avoid the 'dryer up' or worker at the drainer.
(d) Cookers should:
   (i) not be placed under windows or wall cupboards;
   (ii) be placed at least 300 mm away from a corner to allow easy access and door space;
   (iii) be placed in such a position that will prevent another person entering the room and being able to open a door against the cook's back;
   (iv) have an extract fan over the hob (extract fans in kitchens are now covered under the new amendments to the Building Regulations);
   (v) if possible be in line with the sink and not on the opposite side of the room;

(vi) not be separated from the sink by a door – someone could walk into the cook who could then be scalded.
(e) The washing machine should be near the sink.
(f) Tumble driers should be on an outside wall and vented.
(g) Electrics should allow:
    (i) a 30 A control switch behind a free standing cooker.
    (ii) a 13 A switch for each piece of major electrical equipment (e.g. dishwasher, washing machine, refrigerator, tumble dryer). These should be provided at working height and be connected to outlets below the work surfaces.
    (iii) socket outlets to be sited at least 150 mm (6 in) above worktop height or the floor (measured from the bottom of the socket).
    (iv) the cooker control panel to be at the side of the cooker and not above it.
    (v) wherever possible twin socket outlets (this helps to prevent trailing flexes).
    (vi) neon indicators to switch spurs.
    (vii) adequate lighting.

As I have already said, these are ideals that are very rarely achieved but the basic concepts are worth bearing in mind.

Now, down to business.

PRELIMINARY INSPECTION

When carrying out a Building Survey, try to create a routine that works for you.

There is no absolutely standard way of doing a survey. There are times when conditions force you to deviate slightly from the norm but logic would dictate that if the way you operate is systematic then you are less likely to miss something.

As I have intimated before, I like to give the property a quick preliminary overall inspection before I start in earnest and make rough sketches of the internal layout. I find this essential when dealing with a large house. This procedure is useful even when carrying out a survey of a small bungalow.

Remember, you have to be able to understand your own notes and be able to transform that information into a report once you have finished. I find that sketches can be a boon.

In addition to sketches, I photograph each elevation using a Polaroid camera. When I'm actually writing the report, I can then refer to the photographs and check that the written (or verbal) details are accurate, having photographs to refresh my memory. I always take the precaution of putting the date and address and some sort of reference on each photograph so that it is obvious from which direction it is taken (i.e. gable end, front elevation, rear elevation). I always keep these photographs with the survey notes once I have finished with them and they then form part of the permanent record.

NB There are times when either common sense or the client's requirements dictate that photographs should be incorporated into a report. Obviously, as a Polaroid camera takes only 'one off' snaps, if several copies or enlarged photographs are needed for the client then a good quality camera will have to be used.

If you have the time, it is also useful to obtain the estate agents details prior to your visit. These have to be treated with caution but, once again, it could provide useful information concerning the property.

THE INSPECTION GENERALLY

After making my sketches, I normally inspect the house externally first, starting at the top and working down (i.e. roof, chimneys, gutters, rainwater pipes, soil and ventilation pipes and walls, windows and doors). On large properties, I do this elevation by elevation to save walking. On very small properties where it is relatively easy to move around, the whole roof, all elevations, etc., can be dealt with at once.

When recording defects, make sure that the locations are noted accurately. If there is a crack in the brickwork above the back door, say so.

Where it is not possible to inspect a particular item from ground level or off the short ladders carried, make a note to look from inside (if this is possible). For instance, flat roofs over bays that can't be seen off the ladder might be visible from the first floor. It is essential to note all defects found. If a pane of glass is cracked in the front lounge window, record it, and say so in the report.

Once the examination of the roofs, walls, windows, etc., outside has been completed, I usually then go into the roof space. Some people prefer to inspect the inside of the house first, but as the loft usually has a large amount of exposed timber and is the most likely place to find leaks, I like to look there first and then check the findings when inspecting the rest of the property.

I always slip on a set of overalls when I'm about to climb into the loft. This is because, generally speaking, lofts are filthy places. I also ensure that I use a powerful lead lamp to provide light.

Torches are usually too weak to light loft spaces adequately. Access to the loft must be borne in mind. I always carry a set of three metre, three way ladders that can be used as ladders or as stepladders for access to places like lofts. Be careful in lofts and don't take unnecessary risks. Only step on ceiling joists or walk boards.

Don't stand on plastered ceilings, otherwise you could fall through and be seriously injured. This problem can also occur where floorboards are badly rotted. If you have a fall, you could be injured. Even if you are not, the owner will undoubtedly present you with a repair bill.

Make sure that you run your lamp along all exposed timbers that are easily accessible . . . It takes time but it is important not to skimp.

Once the loft has been inspected, wash you hands and ensure that you are clean before proceeding through the rest of the house. *Don't forget, it is polite to ask permission to use the bathroom.*

I usually inspect the top floor next, room by room, making a record of the ceilings, walls, windows, skirtings, doors, floors, electrics, radiators, etc., and then go to the ground floor. This is inspected in the same way. I always carefully inspect the fittings in bathrooms and kitchens to ensure that leaks or the like are noted. When inspecting a wash basin or sink, run your hand along the pipes under the taps. If they are dry then all should be in order. If they are wet, find where the leak is and what damage it has caused (particularly to the floors).

Where possible, links should be established between external faults and those found internally (e.g. leaks found on chimney breast – flashings missing on chimney or a broken rainwater pipe found and damp stains found internally).

If the house has a cellar, this should be inspected as carefully as the roof space as it is likely to have exposed timbers than can easily be inspected for dry/wet rot and woodworm.

Once the house inspection has been completed, the gardens, garages, fences and the like should be inspected.

IMPORTANT ITEMS TO CHECK

Whilst you are looking around the property look out for the following:

(a) Missing tiles/slates.
(b) Flat roofs laid with no falls. (Moss covered.)
(c) Missing flashings or soakers. (See Fig. 4.4.)
(d) Damaged chimney flaunching. (See Fig. 4.4.)
(e) Broken gutters/downpipes. (Any moss apparent on walls?)
(f) Leaking valleys and/or secret gutters. This is especially important around chimneys. Check the moisture reading on the brick stack or the render surrounding the chimney and look out for water runs. If the flashings around the chimney are defective, some sort of stain mark is usually obvious.
(g) Damp patches on walls.
(h) Plumbing leaks.

(i) Condensation in main roof spaces. (Look for 'spotting' on the insulation quilts, particularly beneath laps in the sarking felt.)
(j) Condensation in flat roof voids. (Stains on ceilings.)
(k) A damp proof course that either does not exist or is defective.
(l) That the damp proof course has not been bridged by flags or earth.
(m) Blocked air vents. (*One house that I visited recently had all the airbricks that ventilated the ground floor void blocked off . . . When I made enquiries, I discovered that there had been a plague of mice one year and the owner had covered them up to stop vermin getting through the oversized appertures in the airbricks . . . The covers had never been removed.*)

NOTES ON DAMPNESS GENERALLY

I have detailed in Chapter 12 the problems that can be created by damp penetration.

During the inspection, you must look out for possible problem areas. You have the knowledge, all you have to do is have the confidence to apply what you know.

You have to be on the lookout for signs of water damage/dampness while inspecting the property. Watch out for dark stains on wallpaper, efflorescence (i.e. white powdery patches) on plaster, mould growths or moss.

As you will realize from Chapter 12, if dampness continues for a long period of time then damage will be caused to:

(a) Timber
(b) Plaster
(c) Brickwork.

In timber, excessive damp can lead to outbreaks of wet and/or dry rot as previously discussed. Remember, if it's kept dry then it's safe.

If a damp patch is found, there will be a cause (e.g. a leaking gutter). Try to find the source of the water so that the defect can be put right.

Remember, water runs. So a fault in one part of the property can make its presence felt somewhere else.

Therefore, water dripping down the flex of a light fitting or through a crack in the plasterboard ceiling could indicate many things. There could be a leak from a water tank or a defective flashing in the roof space but the position of the leak could be sited quite a long way from the running water or stain.

In one instance, I know of a case where the insulation in the loft had been incorrectly laid preventing the outside air from ventilating the loft. During a very cold winter, condensation froze on the sarking felt. A layer of ice built up over a period of weeks and when the thaw came, it literally rained inside the loft. The first sign the owners had of problems was when water stains appeared on the ceiling. They, naturally enough, thought that the roof was leaking.

It is also important not to assume that because the roof above a damp stain seems sound, that the repair has been carried out. It will be necessary to examine the roof carefully and try to identify where the roof was leaking. (As I have said before, sometimes it can be a blessing in disguise if it is raining at the time of doing a survey because severe leaks become very obvious.)

When trying to find a fault, apply logic. If the water stain is on a ground floor wall, then defective brickwork or DPC is the most likely cause. A stain on an upstairs ceiling is most likely to be a defective roof.

If the ground floor ceiling is damp, then the most likely cause (if it is well away from a wall) is a leaking pipe or tank.

Sometimes it helps if you quiz the current owner. Domestic accidents are not uncommon and an overflowing bath, while not recommended, is unlikely to cause problems if just a one off incident, whereas a continuous source of dampness is virtually guaranteed to cause some sort of woodrot.

On the ground floor, near the bases of internal and external walls, dampness is almost certainly an indication of rising damp (i.e. a defective damp proof course) if high moisture readings on the skirtings show up on the moisture meter. If the skirting readings are low, the dampness is more than likely caused by condensation.

SIGNS OF DAMP ON UPPER CEILINGS

Any signs of damp on top floor ceilings is likely to indicate roof leaks (i.e. defects in the roof or roof coverings). The most common cause is a slipped/missing flashing/slate/tile or a perished flat roof. It should be possible to locate these defects from ground level using binoculars. Hidden valley and/or secret gutters are more difficult to locate from ground level. Whilst inside the roof space, you must therefore inspect the bottoms of valleys and/or secret gutters very carefully (see Figs 11.25 and 11.26). Another item which can obviously cause damp patches is a defective water tank or leaking pipework. Damp on the landing ceiling directly under the tank probably means a leak on the tank or pipe leading to it. Always inspect inside old steel tanks in loft spaces. These rust through comparatively quickly and always from the inside out.

There are other causes of water penetration such as omission of cavity trays (see Fig. 12.3).

Inbuilt design problems are one of the things which the experienced surveyor homes in on without thinking. After a while it becomes instinctive. He/she knows that certain features in building will eventually cause problems.

DAMP ON GROUND FLOOR/LOWER FLOOR CEILINGS

Make a special inspection beneath flat roofs to extensions. Flat roofs covered with standard BS 747 felt are notorious for developing leaks after only a few years. Standard felt coverings to flat roofs bubble and split and the flashings at abutments come loose and allow water to enter the building. Once again, water stains give away leaks.

Look also for 'cut-off' rainwater pipes that discharge onto flat roofs. It is quite common for people to discharge pipes direct onto flat roofs of extensions forgetting that eventually the continual wearing effect of running water will bore its way through the felt.

Leaking WCs, baths, pipes or central heating pipework can also make themselves apparent on ceilings. If staining is noted, then make sure when checking out WCs that carpets are rolled back. If there is a leak, staining will probably be apparent on the floorboards. It is usually not possible to inspect beneath baths (unless the bath panel can be easily removed), but it should be borne in mind that leaks under baths can cause problems. Leaking fittings must be treated as important because persistent leaks can easily result in wet or dry rot outbreaks.

If you have serious misgivings, you are advised to qualify your report accordingly.

DAMPNESS IN EXTERNAL WALLS ON UPPER FLOORS

In houses having solid walls, perished mortar joints and bad pointing can allow water to penetrate through to the inner face and dampen the plaster and woodwork (see p. 87). In cavity walls, this is normally not possible unless there are mortar 'snots' on the cavity ties which would breach the cavities' defences or allow water to dampen the inner skin of the cavity wall. In solid walls, repointing the brickwork will normally solve the problem but bear in mind that the saturated bricks will take some months to dry out. It is also possible to cover the outer faces of solid walls with clear silicon (after repointing) and this will help to repel rainwater. The other obvious points of weakness in a wall are leaking pipes and gutters. In one survey I carried out recently, I found that the main cause of damp in an upper wall was caused by a rotted timber gutter that was allowing water to leak into the top of the wall. Windows too can allow in water if the mastic pointing is missing or if the cill is not throwing the water clear. Some houses also have projecting band courses or corbels. If the lead flashing above is missing, these can act as a water trap and allow water into the building instead of throwing it clear (see Fig. 2.1). Occasionally, condensation can be a cause of dampness. With the advent of tumble dryers, this has become a bigger problem if the dryers are not vented properly (see Chapter 3). Houses with cavity walls are less likely to have a condensation problem because the construction gives the wall a better thermal value. The most obvious signs of condensation are black or green moulds.

DAMPNESS IN GROUND/LOWER FLOOR WALLS

The ground/lower floor walls can have similar defects as the above but, in addition, rising damp can occur on ground floors. This is usually caused by a defective or non-existent damp proof course or one that has been bridged (see Preface). Flower-beds banked against an outside wall can easily 'bridge' a DPC.

The most common remedy for rising damp is a new injection damp proof course. Silicon solutions are pumped into the bricks (or allowed to soak in under gravity) forming a new barrier against rising water but all reputable DPC specialists recommend that the first metre of plaster be replaced because salts build up in the old plaster and these attract moisture from the air in the house. On a large council estate that I worked on, the local council, in order to save money, replaced all the damp proof courses in the houses but did not carry out replastering works (despite advice from the specialists). Within six months, people on the estate in question were complaining of damaged decorations and black mould. The more vociferous tenants were claiming they still had rising damp and that the specialists had skimped on the work. Replacing the plaster with a suitable product is essential (e.g. waterproof cement and sand or a proprietary renovating plaster). Normal lightweight plaster such as Carlite is not suitable for the purpose.

On a more general note, under the Building Regulations, DPCs must be 150 mm (6 in) minimum above ground level. If they are lower, rainwater can splash over and wet the brickwork above. Any surveyor must give special attention to the height of DPCs. One 'crime' that I have found on a number of surveys is that path levels are raised by the owners and this too can create a 'bridge' which can defeat a DPC.

DAMPNESS IN OLD QUARRY TILE FLOORS

Old quarry tile floors are always likely to let in dampness (see Fig. 11.6) as they do not usually have a damp proof membrane incorporated in the construction.

The only real remedy is for this type of floor to be replaced with a modern floor.

DAMPNESS IN TIMBER GROUND FLOORS

Where timber floors are at ground level, the most likely parts to suffer from damp, especially in older terraced property, are the joist ends where they are built into walls. If ventilation is poor and dampness is present, wet and dry rot outbreaks can be anticipated. If possible, lift a floorboard and inspect the joists.

DAMPNESS IN CELLARS

Damp in cellars is virtually guaranteed in my experience.

The only real remedy is to 'retank' the cellar in asphalt or alternatively use a Sika or similar render internally to both floors and walls. Make sure that ends of joists built into walls are tested with a probe and moisture meter.

DAMPNESS IN BAY WINDOWS

In older properties it was very common to find that the bay windows had been built in half brick construction (i.e. 4½ in thick, or 112.5 mm). Because half brick walls do not have the same thermal resistance as one brick (9 in, or 225 mm) solid walls or cavity walls, the walls of bays tend to be colder than the rest of the house. Bay windows can therefore be condensation traps and black and green mould can form on the wallpaper. Perished plaster can also result because half brick walls are not very resistant to water penetration anyway. It should be anticipated that bays will be damp.

INSPECTING THE MAJOR ELEMENTS OF THE STRUCTURE

Dampness is, in my opinion, one of the major items to check out when inspecting a property because, as I have indicated in Chapter 12, the timberwork of any property is the softest and most easily destroyed element but there are obviously other defects and these are covered below.

THE ROOFS EXTERNALLY

See Figs 11.21–11.27 for technical terms.

ROOFS GENERALLY

If a roof has been relaid with a heavier tile than the original, and overloaded, the resultant effect can be 'roof spread' (i.e. the feet of the rafters start to move away from the wall plates).

Externally, look at the gutters and the ridge line. If the ridge has a 'bow' in it and the gutters have obviously been displaced, then the roof could be spreading (and might eventually collapse).

Another sign of 'spread' is when struts in the loft space show signs of bending or have snapped. If purlins are bending, then they are probably overloaded or defective.

Look out for moss as opposed to lichen. Large amounts of moss give a strong indication that roof tiles have started to degrade and that the upper surfaces are perishing.

Missing ridge tiles should be checked out. The same goes for hip tiles. As indicated in the sketches, ridge and hip tiles are purpose-made tiles designed to cover the sharp angles formed at the ridge and hips. These tiles are bedded and pointed in mortar. You should carefully inspect this mortar through binoculars to see if it is missing. If it is, it needs replacing. Cracked tiles or slates should also be noted.

Slate roofs have their own problems. Although the slates will last almost twice as long as tiles (i.e. about 100 years), the nails that hold them in place tend to corrode and the roof becomes 'tired'. Roofs that have become 'tired' can usually be noted very easily. Look out for slipped slates and slates fixed with 'tingles' (a tingle is a bent piece of metal slipped under the slate to support it). If slates are tingled then it doesn't take much intelligence to realize that probably the whole roof needs stripping and renailing.

When inspecting roofs, look out for slipped flashings, aprons and other leadwork. (See Figs 4.4 and 11.25 in particular.)

Bays and porches quite often have a flat roof covered in lead or three layer felt. Flat roofs should not be totally flat, but should be built with slight slope. If possible look at the top of flat roofs. There could be a layer of limestone chippings on the upper surface of felt covered roofs and there should not be any standing water on any type of flat roof.

Then there are irregularities in the roofing generally. An exterior inspection might lead you to suspect a roof defect, but only an interior inspection will properly reveal if there is one. If the cause of the defect is woodrot, the roof covering can be stripped back, the defective timber replaced with treated timbers and the coverings replaced. Undersized rafters can sometimes be reinforced with additional purlins, struts, etc. However, any roof alterations of this nature need to be carried out by an experienced contractor.

Inspecting the roofs of neighbouring properties can also be useful. Roof tiles have a long life (say, approximately 60 years). As houses tend to be built in groups or estates, if the roof tiles of several neighbouring houses have been replaced, then the chances are that the roof of the property that you are inspecting may be in need of renewal.

When inspecting the external surfaces take particular note of any broken, slipped or missing tiles or slates. There are two reasons for this. The first is that they need replacing; the second is that if there have been leaks, then whilst inspecting inside the loft, you must look for damage (i.e. outbreaks of rot or other water damage).

In the UK, most domestic properties have tiled or slate covered main roofs (pitched roofs). I have heard tales of surveyors who have inspected pitched roofs by walking on them but, in my opinion, it is an unsafe practice and damage can actually be caused to brittle tiles or slates by standing on them. Roofs are not designed to walk on and two storey domestic properties can usually be inspected from ground level.

When viewing a pitched roof of a house, it is best to stand well back so that the whole slope is visible and then study the roof surface by sight or with the aid of low power binoculars. Sometimes, in order to obtain a good view, it is necessary to go across a road.

CHECKLIST

Look out for the following:

(a) The ridge should be level and ridge tiles should not be missing. A dip in the ridge could be an indication of trouble.
(b) There should be no sags noticeable in the roof coverings.

If there are any dips or troughs, then it could be caused by:

   (i) Corrosion of plates or trussed rafters.
   (ii) Corrosion of nails, bolts or connectors.
   (iii) The roof being recovered with heavy tiles and the timbers being overstressed.

(c) Note cracked, slipped or missing tiles or slates. Also look for tingles on slate roofs.
(d) Note if a pitched roof has had felt applied to the surface? If it has, it is a cheap way of waterproofing an old defective roof. This type of treatment does not last for long. Sooner or later, someone, possibly the person that you are doing the survey for, will have to 'pick up the tab' for completely re-roofing the house at some later date. Also, as with 'The Wetlands' this type of covering can create condensation inside the loft if ventilator tiles have not been inserted (see Fig. 4.8).
(e) Check out flat roofs to extensions and bay roofs very carefully. Look for cracks and bubbles in the surface. BS 747 felt roofs should be covered in white limestone chippings bedded in bitumen. Quite often the chippings have been washed away and are lying in the gutters.
(f) Check roofs to dormers, which are particularly prone to leaks. Look out for stains on dormer ceilings once inside.
(g) Look for obvious signs of cavity-wall tie lift where roofs meet gable ends (see p. 106).

CHIMNEYS

As chimneys are one of the most exposed parts of any house, they need repointing more often than the main brickwork. Defective pointing can be easily noted through binoculars. TV aerials that are badly fixed can also damage stacks.

Around the base of the chimney pots there is a cement mortar capping called the flaunching. This breaks up and degrades after a few years. Once again, it is usually possible to note the defect from ground level. Where flaunchings are defective, these can be hacked off and replaced.

Around chimneys (see Fig. 4.4) where the stack penetrates the roof, the gap is filled with flashings and soakers. These are purpose made pieces of lead or lead substitute. The flashings are stepped on the sloping sides with aprons on the non-sloping sections. Missing leadwork must be noted because any defect in the lead will let water into the roof space.

Look out for cement fillets around chimneys – fillets of this nature are bad building practice and become defective very quickly.

The only reason for a chimney existing is that it discharges the waste (smoke) out at high level where the wind can't blow it back through the windows. A chimney may have several flues. Chimney stacks should not lean in any direction but leans are not uncommon.

One of the possible causes of a leaning stack can be a sulphate attack. The sequence of events is as follows:

(a) Driving rain saturates a chimney stack for a period of years. The north side of the stack normally doesn't dry out as quickly as the sunny southern side and the continuing dampness carries soluble sulphates in the bricks or chemicals in the flue gases to the cement in the mortar. These then react, causing expansion on one side of the chimney.

(b) As the attack proceeds, the bond is lost by the mortar on the damp side and the stack bends and finally leans one way (see Fig. 4.3).

The only lasting cure for a leaning chimney is to demolish it and rebuild the stack. When rebuilding, the stack should be constructed of special quality bricks to BS 3921 and/or sulphate resisting Portland cement used in the mortar.

Another cause of leaning stacks is a missing chimney breast. It is not uncommon for people to demolish a lower chimney breast without putting in RSJs (rolled steel joists) to support the structure above. Be on the look out for missing chimney breasts below stacks.

Within the roof space, it is normal for the chimney to be rendered. If it isn't, say so in your report.

THE EXTERNAL WALLS

(See Figs 4.7 and 11.9–11.11 for details of brick bonding)

I have covered the details of brick bonding in 'The Wetlands' (p. 26). In newer property, most walls are built in stretcher bond and indicate that it is probably of cavity construction. The presence of 'headers' indicates a solid wall construction that is less water resisting.

Walls should be vertical. Check the corners with a long spirit level. Are they vertical? There should not be any bulges or cracks.

Cracks in a wall indicate several defects. Where the house is constructed of brick, cracks tend to follow the horizontal and vertical joints zig-zagging down the mortar beds in a roughly stepped diagonal line.

Settlement and cracking can occur if a large tree is growing near to a house. Trees can extract large amounts of water out of the ground and this will cause clay soils in particular to shrink (see Fig. 12.10).

It is also worth remembering that a small insignificant tree planted now can grow to a massive size in a few years. A Lombardy poplar could easily reach a height of 30 m when it is mature. The roots can extend for a considerable distance. Positioning of trees does bear some thinking about.

A leaking drain washing soil from around the house foundations can create a similar effect.

Another cause of ground movement can be caused by mining subsidence. By mining, I am also including any extraction process. In places like Northwich where salt is extracted from the ground, subsidence is fairly common. There are other areas that I know of where houses have been built on sand. Because the foundation design is incorrect, settlement is a general problem.

If settlement of any major proportion has taken place in a building, the only real cure is to demolish the building or to underpin the old foundations.

The underpinning process comprises digging out under the existing foundation (in short lengths – usually of not more than 1 m). In order to get at the base of the foundation, a preliminary trench is dug to the level of the original foundation.

The existing foundation is then undermined, the void filled with concrete and any gap between the new and old concrete packed with a strong cement grout to ensure that the loads are transferred safely to the new concrete and thence to the ground.

Underpinning is a very expensive operation and not one to be undertaken lightly. However, it is usually cheaper than complete demolition. Cracking of walls can also be caused by cavity wall tie corrosion (see Fig. 12.6 and Chapter 12).

JUNCTIONS BETWEEN THE MAIN HOUSE AND EXTENSIONS

(See Figs 12.1 and 12.2)

Junctions between extensions and the existing property should be 'cut toothed and bonded' to the existing wall every other course (i.e. bricks let into the old structure to ensure continuity). Some builders just butt up the new brickwork to the exisiting. This is bad practice and creates a weakness between the two structures. The only exception to this is when patent continuity strips (e.g. Furfix) has been bolted to the old brickwork so that the new walls are tied in with the existing ones. The only way to check if this is the case is to inspect the plans if they are available.

DAMP PROOF COURSE

See previous notes concerning DPC.

AIRBRICKS

Where there is a suspended timber floor in the ground floor, there should be $1500\,mm^2$ per metre run of external wall or $500\,mm^2/m^2$ of floor area whichever is greater (see Fig. 11.7).

RAINWATER GOODS (GUTTERS AND DOWNPIPES)

Gutters and down pipes (see Figs 4.6, 11.23 and 12.5 for details) are made from a variety of materials (i.e. wood, uPVC (plastic), cast iron, asbestos, cement and aluminium).

Gutters and downpipes on modern properties are usually made of uPVC (plastic).

Plastic guttering is reasonably maintenance free but will not take the weight of ladders and will deform if maltreated.

On older properties you must expect to find any of the following:

(a) Cast iron
(b) Timber
(c) Timber lined with sheet lead.

In theory, with good maintenance, most of the materials listed (with the exception of timber) should have a fairly long life but because of lack of maintenance (cleaning and painting) leaks and decay can be anticipated.

Check gutter and pipe joints. After a downpour is best because leaks tend to persist for several hours after it has stopped raining. Check walls and flagged paths near pipes for moss growth. This could indicate a leak. Check the gutter brackets. If they are rusted away, the gutter will sag.

As I have already said, cast iron rusts in areas where it is difficult to paint (i.e. the backs of pipes and under eaves).

(If you refer to most of the reports in Appendices A–E, faults were found on the gutters and downpipes – note the wooden gutter on Fern Lea.)

Gutters will also leak if they are laid to incorrect falls.

It is normal for the sarking felt to be turned into the back of the gutter (see Fig. 11.23). On one survey that I carried out recently, this had not been done and as the roof tiles did not project over the gutter, water was running behind the gutter and not into it. Evidence of the spillages was there to be seen. There was a great deal of moss on the path and water stains were apparent on the fascia board. Some houses have secret gutters behind parapet walls. As with any other secret gutter (see Fig. 11.25), leaks can occur without them becoming immediately apparent. In houses with parapets, check tops of walls and ceilings directly below the gutter with a moisture meter.

SOIL AND VENTILATION PIPES

(Refer to Fig. 12.5)

For further information refer to The Building Regulations (Approved Document H).

The soil and ventilation pipes (SV&P) are the large pipes that take the waste away from the toilet (WC), bath, etc., and connect to the drains. On older houses these are cast iron and on newer houses uPVC. *Some people refer to them as 'stink pipes'.* My previous comments on inspecting the backs of any pipes for corrosion still apply but there are other considerations.

As anyone knows who has studied physics at school or has ever attempted home made wine, if you want to get liquid out of one vessel into another, it is possible to make a simple syphon with a length of tube. Once the syphonage has started, the liquid will empty into the second vessel without any problem as the weight of the water descending in the tube will suck the liquid out of the first vessel.

As you will probably be aware, most sanitary fittings (toilets, wash hand basins and sinks) have a 'trap' underneath. This holds water which prevents foul smells entering the dwelling from the drainage system.

The principle of syphonage caused a problem for plumbers because sometimes on a long run of pipe, syphonage would start up and empty every trap in the house thus defeating them.

This problem was overcome originally by putting in additional ventilation pipes, the sole purpose of these being to let air into the system and prevent syphonage occurring.

If one looks at the back of one or two older buildings in your town, it is still possible to see examples of 'two pipe' systems.

The problem of syphonage has largely been overcome nowadays by a great deal of research. The 'single stack' system came into existence some years ago and the technology is utilized by all the major plumbing manufacturers (e.g. Terrain, Marley, Hunter).

However, the single stack system works by very specific rules and unless these are observed the system will fail.

It is therefore essential that you are on the lookout for DIY plumbing and obvious examples of disregard for these rules. With the one pipe system the length of a pipe is strictly governed by its diameter. The table below gives simplified details.

Pipe diameter	Length (max.)
30 mm diameter pipes	1.7 m
40 mm diameter pipes	3.0 m
50 mm diameter pipes	4.0 m
WC pipe	15.0 m

If you refer to Fig. 12.5, you will note that there is an extra long waste pipe that has been installed by the present owner. The cause of the waste pipe was the desire to have a vanity unit in one of the bedrooms. Because it is so long, it will more than likely start to syphon off the water in the trap every time the bowl in the vanity unit is used and foul smells and gases from the underground drains will pervade the bedrooms.

EXTERNAL WOODWORK ETC.

You should carefully check all doors and windows externally. Softwood windows in most modern houses are of very poor quality. One can virtually guarantee woodrot in untreated windows once they have been in a modern property for a few years. If paint work is in a poor condition, give the woodwork extra special attention. High level woodwork such as fascia boards and soffit boards (Fig. 11.23) can be a problem to inspect but visual inspection, with or without the use of binoculars, should give a reasonable indication of their state of repair.

Obviously, doors and ground floor windows are easier to get at. Check the timber of cills and junctions of timbers where water lodges within frames, for soft spots. Tests with a probe and moisture meter are useful. Make sure that there is no missing putty in doors and windows. If water does lodge behind putty, it can cause rot inside frames. Localized woodrot can be spliced or repaired with a patent filler such as Tetrion. Badly rotted windows and doors may need complete replacement. Cills of windows are generally the first to become defective. Make sure that you test cills carefully. It is also worth looking under cills to ensure that the hidden timber has been well painted (remember out of sight can mean out of mind). The underside of the cill should also have a groove called a drip in it to prevent water running against the wall of the house.

Sometimes houses have sand stone cills. If possible, check them with a spirit level. Remember, it is reasonable to assume that the cills were laid level when the house was built. Cills out of level could indicate subsidence. Look for cracks and missing stonework in cills. This again can indicate settlement but cracks and missing stone can also trap and/or divert water in the building.

Similar checks should be carried out on metal windows as on timber windows, only there the major defect will be rust as opposed to woodrot.

ROOF SPACES

(See Figs 11.21 and 11.22)

In my experience, the roof space is one of the most useful areas of the house to inspect because timberwork and the underside of the roof are usually open to view and the effects of water penetration can usually be detected easily.

I have already stressed the dangers of loft spaces. While you are moving around, be careful. Test each step, ensuring that you are holding on to main timbers as you move.

It is totally unsatisfactory for the surveyor to open the loft access hatch and merely shine the torch or lead lamp around the loft area.

Doing so is pure laziness and will inevitably get you into trouble.

You *must* climb into the loft void and inspect *all* the timbers that practicably can be checked. The lamp must be passed up and down each timber in turn and you must look for signs of woodworm and rot. Moisture tests must be taken on the timbers and the percentages noted. If they are found to be in the 'green zone', then the likelihood of there being any rot is very low. (See Chapter 15 for details on the use of a moisture meter.)

Find the chimney stacks and pay special attention to the timbers around them and take moisture readings on the brickwork or render around the chimneys. Look for leaks or stains on the chimney itself. High readings probably mean that the flashings and soakers are not totally effective (see Fig. 4.4).

After saying all that, sometimes it is physically impossible to inspect the entire roof space.

As an example of this, I recall a loft which was in two halves. There was a wall that divided the two parts. The access hole (approximately 12 in (300 mm) square) between the two was so small that it was impossible for me to get my shoulders through the gap. I tried to see as much as possible through the hole but recorded in the report the lack of access and the fact that I could not inspect the second loft area. I also advised that the access hole be enlarged.

One system that I have found useful when inspecting roof spaces (after I have found a safe position) is to turn off the torch or lead lamp and look for daylight especially around chimneys and other perforations in the roof covering (e.g. around soil stacks).

In modern buildings the underside of the tiles or slates will be covered by felt (sarking felt or underfelt). This is placed under the tiles or slates to act as a secondary barrier in case a slate or tile is dislodged. Unfortunately, some people rely on this secondary barrier instead of making sure that the main defence system (i.e. tiles or slates) is sound.

When inspecting the loft space, I would suggest that the following checklist be followed:

(a) Can you see the tiles/slates (if not, see (b), below). Or is there a sarking felt underneath?

(b) The main roof area – are there any tiles or slates missing? If slates and/or tiles were noted as being missing during the external inspection, try to locate the areas whilst inside the loft. Timbers around the area should be tested with a moisture meter.

(c) Is water obvious around the chimney or soil and vent pipe (SV&P)? (If you get a high reading, then a leak is fairly certain.)

   Check under valleys. If the lead in the valley is defective, the moisture readings in the valley rafter will probably be high. (Out of the 'green zone'.)

(d) Check sides of chimney stacks. The chimney should be rendered where it passes through the loft. Make a note as to whether or not the render is there. Check the chimney for leaks. If the stack has obvious water runs down the side or is very damp, there is little room for doubt; there are leaks on the flashings above or the chimney itself is letting in water.

(e) Timbers should be checked for large splits or twists. Examine all timbers carefully for signs of woodworm and woodrot.

(f) Ensure that you check the internal gable walls between semi-detached properties and terraced properties. It is common for bricks to be missing or for there to be access holes into the neighbouring lofts. These perforations present a fire risk. If the next door house were to go up in flames, the fire could work its way down the street.

(g) Inspect the cold water storage tank and header tank. The tanks should be lagged and have an insulated cover. All pipes should be lagged, new PVC tanks are better than galvanized steel tanks because they don't rust. All tanks should be inspected for splits or leaks. Look around the base of the tank for possible leaks.

(h) Inspect the thickness of the insulation to the ceiling. The 1990/92 amendments of the Building Regulations require that insulation should provide a 'U' value of 0.35. In practice this means that a glassfibre quilt of 150 mm (6 in) will be required. Most lofts will not have this thickness of insulation unless the owner has 'upgraded' the insulation standards to suit the modern regulations.

   Insulation should be laid between the joists and not over them. Look for drip marks on the quilt. Drips below laps in the sarking felt can indicate condensation problems.

(i) Check electric cables. Rubber insulated cables have a limited life and indicate that the house will need rewiring. PVC cabling indicates that the house probably has a fairly modern electrical installation.

INSPECTING ROOMS

Internal walls, ceilings, plaster and decorations

The condition of ceiling plaster, wall plaster and decorations does not in itself affect the basic structure of the house. However, cracks, bulges, stains, etc., can indicate problems below. Lath and plaster ceilings, and lath and plaster to partition walls, were common before the Second World War.

These should be given careful inspection where possible because, as previously mentioned, dry rot will travel, and the buried laths create an interface between different parts of a house. Don't be confused by hair cracks. Plaster cracks over 3 mm will probably indicate that all is not well. Steps in cornices could well indicate subsidence. It should be borne in mind though that it is very common for cracks to occur in plasterboard ceilings and at junctions with walls. These cracks however run in straight lines, whereas other cracks tend to be irregular.

Lath and plasterwork does deteriorate over the years. Once this happens, bulges will appear in plaster. Sometimes old ceilings have to be hacked down and replaced. In other instances, **overplating** can be the solution, fixing plasterboard to the old ceiling and using extra long nails.

Beware of ceilings and walls that are covered with polystyrene tiles, timber boarding or old ceilings that have been Artexed, especially if some rooms are Artexed and some are tiled. These are obvious signs of defects being covered.

Around chimneys, test the plaster to ceilings and walls with a moisture meter, especially on upper floors as it is a good check on defective flashings and excess moisture from chimney flues. All internal walls should be vertical. Tap walls: if the plaster rings it could mean that the wall plaster is 'boxy' (i.e. coming loose). Tapping walls can also help to find stud partitions. These are non-load-bearing and built from timber framework that runs both ways. Onto this framework plasterboard is nailed, scrimmed with hessian and skimmed. Wall plaster should be checked for dampness and excessive cracking. Once again, anticipate finding hair cracks, but wide cracks, especially at junctions, could indicate collapse or subsidence. Areas which are likely to be undecorated (e.g. inside cupboards, understairs, etc.) should be given special attention. Diagonal cracks from doors and windows indicate subsidence. However, hair cracks should not be a source of worry. It is quite normal in modern plaster.

Solid load-bearing walls will sound dull if thumped. The location of internal structural walls is important because internal structural walls act as buttresses for the external walls. Unless the upstairs walls are lightweight or studded partitions, the upper walls should be built off downstairs walls, or there should be beams to support same. Damp to internal and external ground floor walls above the skirting will probably mean that the damp proofing course in the walls is defective. Dampness around windows will probably be caused by condensation. Black and green mould patches are a good indication if condensation is the cause of dampness. When dealing with damp patches anywhere on walls, try to find a loose piece of wallpaper (do not tear wallpaper or the owner will have a justifiable grievance over damage caused) and test the plaster behind. Sometimes a layer of aluminium foil will be found. Using aluminium foil to try to disguise rising damp is a common practice. This treatment is largely ineffective and indicates a long-standing problem.

It is worth checking chimney breasts. If there are more chimney breasts upstairs than downstairs, *beware*. I have seen cases where unknowing owners demolish downstairs chimney breasts without putting in supports. The same thing can occur upstairs, leaving the stack unsupported.

Floors

Wherever possible, turn back carpets and floor coverings. (If it is not possible, record which rooms have fixed floor coverings.)

Obviously, floors should be level and flat. If in doubt check the floor with a large spirit level. A sloping floor is a sign of possible settlement. (For ground floor construction see Figs 11.1–11.7). Suspended floors comprise timber boards or chipboard laid over joists whereas solid floors in modern houses are concrete, but note my previous comments on old quarry tile floors (see p. 98).

Most upper floors are of a suspended construction.

Excessive movements in a suspended floor indicate that either the members are undersized or that rot has affected them. It is always useful to bounce up and down on a floor and check how firm it is. If it is possible, lifting a floorboard is advisable, however, fitted carpets or adhered vinyl sheet flooring often prevent this. If it is possible to get a board up (one near an external wall is best), shine a powerful torch or mains lamp into the void. Anticipate finding a lot of dirt, builders always sweep their rubbish into void spaces. Check joist ends with a probe and moisture meter. If it is dry, then it will be safe. Check the airflow in the floor. If it is possible to feel currents of air flowing in the floor space, then the ventilation is good. Check the airbricks to make sure that they are not blocked.

Solid ground floors should be checked for sulphate attack (see previous notes). Take moisture meter readings on the skirting. Where solid floors have a 'Marley' tile covering, check the floor for dips in the tiles. If dips are observed, then the floor has settled.

INSPECTING JOINERY/METALWORK/STONE CILLS

These should be inspected systematically on a room by room basis. Open windows, inspect the frames closely, look at the cill board. On upper windows check as much of the external frame and cill as possible from inside and by leaning out. Use of a metal probe at intersections of members is advised. Keep your eyes open for any fungal growth or woodworm. Check all window ironmongery to ensure that it is not broken. Make sure that sliding sash windows run easily. Ensure that there are no draughts or obvious leaks or cracked panes of glass.

Check the skirtings and picture rails if applicable. Any deterioration of the wood or irregularities on the paint should be checked. Remember, one of the first signs of active dry rot is waviness in the woodwork. Check the gaps behind these timbers. If the gap is very wide (over ¼ in, 6 mm) then it could be a sign that the floor has moved (smaller gaps would just be timber shrinkage). In older houses, check skirtings in bay windows very closely. Damp penetration in bays is not uncommon.

If in any areas timber feels soft, test it with a probe. It is not advisable to do this in front of the owner. If the probe slides in easily then the wood will undoubtedly be defective. Check that door frames and architraves around doors have upright jambs and level heads. Frames out of square or binding doors could well provide evidence of settlement. Doors should be checked to see that they are square and have not been planed down. If they have, it is possible that they have been adjusted to allow for past settlement. Ironmongery should be obviously working correctly.

Test staircases by walking up and down these several times. Creaking should be investigated. Try to look under cupboards beneath staircases. As the underside of staircases tend to be unpainted areas, check for woodworm outbreaks and also rot where the wall string is bedded in the wall plaster. As with the skirting, check for gaps along the top edge of the wall string. Wide gaps can be signs that walls are moving due to lack of lateral support in the staircase area. Make sure that handrails are safe. All staircases should have handrails and banisters.

THE ELECTRICAL SERVICES

The electrical services are not normally checked in detail by the surveyor, but this does not mean that you cannot make some reasonable assessment of the service. With a house of any age, it is sensible to ask if the house has been rewired. If it has, the present owner can usually provide proof. If this proof is not available, then you must check the system as best you can.

Look at the fittings. If they are modern square pin sockets, then it is likely that the system is reasonably modern. However, if the sockets are found to be round pin type and of various sizes, then it is highly likely that the system is an out of date system which does not comply with the modern IEE Regulations. (The IEE Regulations are a set of rules produced by The Institute of Electrical Engineers.)

Then look for the obvious signs of age, broken fittings, black or brown overheating stains and old rubber insulated cables. Modern fuse boards are reasonably compact and neatly designed. Older fuse boards tend to be very large with black bakelite casings. Bathrooms should not have power supplies or internal switches unless they are cord operated ceiling types.

Check if any fittings have bits missing. On one survey, I found the electrical system to be very unsafe because several of the old switches had bits broken off the covers. Anyone unaware of the defect, or forgetful, could have easily been electrocuted.

The intensive use of extension leads is also an indication that the present system is overloaded.

If the electrical system looks defective, you should have no hesitation in recommending that a qualified electrician test the system fully.

As indicated in the section on roof spaces, wiring is very often on display in the loft. If the wiring is PVC, then the system could be reasonably new. If it is rubber, then a rewire will be needed.

There are sometimes other give-away indications concerning the state of the electrical system. In one property that I visited, the owner had placed a note on one of the sockets which warned the user not to plug in any high power appliances. The meters and distribution system were antiquated and upstairs there were no sockets. You might not be a qualified electrician, but you can use your intelligence. As an example of this, on a survey that I carried out recently when I inspected the meters, the supply authority had helpfully put a label on the meter which gave the date of installation.

You will note that in my reports, I state that the electrical system has not been tested but that I usually schedule the points. This gives the prospective owner a good indication of the usefulness of the installation. A kitchen with only one single socket outlet is obviously grossly substandard, whereas one with four or five doubles allows the owner to fully utilize modern appliances.

GAS SERVICE

Be on the lookout for very old appliances and the tell-tale smell of gas.

The first question should be, is there a gas supply? Some houses do not have this service laid on. If there is, then testing the service is beyond the scope of a normal building surveyor and he/she should note that no test has been made on the system.

PLUMBING AND HEATING

Check the inside of baths, basins and WCs for cracks. Make sure that the traps on WCs are not leaking or broken. Make sure that you turn on all taps and flush all WCs.

During one survey, I noted a broken soil pipe that ran across a kitchen ceiling. The crack was directly above the cooker. It doesn't take a great deal of imagination to realize that foul water could have been discharging into cooking vessels on the stove.

Central heating is becoming the norm for most houses but there are still a large number of properties in existence which rely on night storage radiators, individual gas appliances and sometimes paraffin heaters. Let us assume that the house in question has central heating. As you go round the property make sure that you check each radiator for signs of leakage and/or rusting. You also need to check on the copper cylinder that usually holds the hot water.

The central heating systems should be checked for leaks and 'pinholes' on the surfaces of radiators. Check that the valves operate smoothly. The boiler should be inspected for some indication of capacity. This can then be checked with a heating engineer, if considered necessary. The condition of the boiler is also important. Look for leaks, water stains, or cracks. Enquiries with the owner should be made about the servicing. Gas appliances should be serviced annually. Check out the water tank and cylinder for leaks, rust or corrosion. Check the insulation. The copper cylinder should have a jacket.

DRAINAGE

If possible, lift manhole covers and then ask the owners to run the taps and flush the WCs while you are outside and ensure that the water runs freely in the drains.

You must not assume that the drains are in good condition (or if you do, you must qualify your report . . . No manholes could be found so it is not possible to comment on the drainage system).

On one property that I visited, when I lifted the manhole cover I found water standing within 6 in (150 mm) of the cover. As the manhole proved to be 6 ft (1800 mm) deep, you can see how bad the blockage was. The owner did not realize that he had a problem. (As it happened, he immediately called in a builder and the drains were rodded.)

GARDENS/FENCES/OUTBUILDINGS

General comments should be included on the fences, the state of paths, sheds, garage and the like. Be on the look-out for collapsing boundary walls. Also note any trees growing too near to a property, including those in adjoining gardens (see p. 109).

If you refer to Appendices A–F, you will see the sort of general information which I normally include in a report.

14 Planning and Building Control in England and Wales

WHY INCLUDE A CHAPTER ON PLANNING AND BUILDING CONTROL?

I carried out a survey recently on a terraced property. It was obvious that the roof had been replaced because the surrounding buildings had slate roofs and this particular property had a tiled roof. When I checked with my records for the area, I discovered that it was in a conservation area.

The roof should not have been replaced without permission. As far as I am aware, the council did not know that the work had been carried out. What if they suddenly did find out about the illegal alteration?

Obviously, the people who eventually purchase the property could well find themselves being harassed by the Local Authority. They could have been faced with the cost of restoring the roof to its original condition. I doubt if the average solicitor would have been aware of, or discovered, the problem.

(Solicitors do not normally visit a property.)

On another occasion, whilst carrying out a survey, I came to the conclusion that the work would not comply with current regulations. When taxed, the current owners admitted that they had extensively altered the original structure without consulting the Local Authority.

During that survey, I discovered a whole series of wrongdoings. A loft conversion had been carried out without the joists in what had been the roof space being enlarged or strengthened.

It was easy to check the joist sizes in the remaining loft space and against those listed in the relevant Building Control tables.

The new staircase that served the upper floor did not comply with the regulations, being far too steep and having wrongly sized treads and risers, and a rear extension had been built without any plans being submitted to either the Planning Department or Building Control.

It should be fairly obvious that a surveyor/building consultant who wishes to become involved in Building Surveys should have a good knowledge of both Planning and Building Control matters and that is why I have included a brief chapter on the subject.

WHAT IS THE DIFFERENCE BETWEEN PLANNING AND BUILDING CONTROL?

The easily understood answer is that the Planning Department of any council is interested in what a building looks like, how it will fit into the surrounding area, car parking facilities and the like. In other words, they are concerned with environmental issues and the impact that the building will have on the surrounding area.

Building Control are interested in the structure of the building itself and apply Building Control Regulations to any building that comes within their remit.

These regulations change from time to time. The changes can be caused by natural progression or the need to conserve energy (as with the 1990/92 amendments to the 1985 regulations) or because of a disaster. (Major fires in hotels always tend to make governments of any political persuasion consider the adequacy of the regulations in force.)

Some structures are not subject to the Building Regulations (e.g. carports) but most domestic structures are.

WHEN IS PLANNING PERMISSION REQUIRED?

Most alterations to a house require planning permission and it is relatively easy to check if, say, an extension has approval.

There are exceptions to the rule though. Certain works are 'permitted' as long as the owners comply with the relevant criteria contained within the General Development Order in force at the time that the works were carried out.

The Permitted Development (PD) rules do not apply in National Parks, conservation areas or to listed buildings. So, if you know that the house in question is subject to rules governing National Parks etc., then you also know that planning permission is needed.

The PD rules change from time to time so it is essential that you keep abreast of current regulations. (I have covered PD in greater depth in my book *A Practical Guide to Single Storey House Extensions*.)

In simplistic terms, if an extension is at the rear of a property and does not exceed $50\,m^3$ on a terraced house or $70\,m^3$ on a semi or detached property, then it will probably be permitted as far as Planning are concerned but the Building Control regulations are *not* covered by PD. Any habitable room must comply with the Building Regulations and be approved by the council.

(Beware . . . If the property has a road at the front and back of it, then PD is probably out of the question. NB Porches have their own PD rules.)

However, the *interpretation* of what constitutes PD, in my experience, varies from council to council and if in doubt you must check.

WHEN IS BUILDING CONTROL APPROVAL REQUIRED?

The simple answer is that extensions to most habitable rooms require Building Control approval. Subject to size restrictions, conservatories, covered ways and porches are now no longer subject to Building Control supervision.

HOW DOES THIS AFFECT A BUILDING SURVEY?

The sensible surveyor will ask the current owners or their agents to provide proof that any recent alterations/extensions have Planning approval/PD status and Building Control approval.

Sensible owners will have used the services of a surveyor or architect when they built their extension and will have retained copies of the relevant approvals and/or letters from the council. All Planning and Building Control approvals have a unique reference and these can be checked out.

Very often, solicitors will ask for these documents but as I have said before, solicitors do not usually carry out site inspections and may not be aware that the property has been extended. To be sure, if someone has done something illegal, it is unlikely that they will volunteer the information. It is one more reason for a prospective owner to have a Building Survey carried out.

LISTED BUILDINGS/CONSERVATION AREAS/TREE PRESERVATION ORDERS

Listed buildings

If a building is **listed**, it means that demolition, alteration and additions will only be allowed after the Local Authority have carefully examined the alterations or extensions to ensure that the proposals do not deface the character of the original.

Listing of buildings does not just affect stately homes. I have known instances where humble stone built terraces have been affected, and quite rightly too, in my opinion. There are too many instances where beautiful buildings that have survived for hundreds of years are destroyed forever just because of the thoughtlessness or ignorance of one owner. The attitude *'it's mine, so I can do what I like'* is still too prevalent, despite the fact that more people are now shouldering their responsibility to future generations.

If for no other reason, the tourist trade is becoming a major source of foreign currency, and if we allow fools to destroy our heritage, then there will be nothing for tourists to see . . . If there is nothing to see then they won't come to the UK.

Demolition or alteration of a listed building without permission is a punishable offence.

Conservation areas

A conservation area can cover a whole village, the centre of a town, a terrace or a small group of buildings.

Tree preservation

A Local Authority can make a tree preservation order relating to any tree, group of trees or a belt of woodland in gardens, building sites or fields. Preserved trees cannot be interfered with, without permission. (This includes pruning branches and roots.)

It is one of the reasons why I always qualify my reports when I recommend that a tree be removed.

Generally

The prospective owner of a property will obviously want to know if the building that he/she is thinking of buying is listed/in a conservation area/has tree preservation orders in force.

If you are unable to provide this information in your report, then it is advisable to say so, suggest that your client asks his solicitor to make these specific enquiries. If you are aware of any or all of these factors, then they must be included in your report. (*And qualified if necessary.*)

I NEED MORE INFORMATION

Can a surveyor be expected to know *everything*? The answer is obviously 'no', but he/she should try to keep abreast of the latest changes in the law and regulations that affect buildings.

If you know enough, then you understand one of the most important lessons of life. You start to realize just how little you really do know. However, help is at hand. Information is everywhere and most of it is either free or relatively inexpensive.

Have you bought a copy of the latest *The Building Regulations 1985 and 1990/92 amendments*? The regulations and approved documents contain an enormous amount of information. You are strongly recommended to study these documents (if you have not done so already).

(See Chapter 18, p. 167, for other useful texts.)

Once you are familiar with the new regulations, even if you cannot recall every detail, you know where to look for the information that you require.

The Planning Departments of most Local Authorities have guidelines and documentation that they hand out, free of cost. I would suggest that it makes sense to avail yourself of this free information, and study it carefully. If you are aware of local conservation areas and the like, if you know where a new motorway link is proposed and if you know the recommended limits of the sizes of extensions, then you are a few steps ahead of the game.

If all else fails and specific information is required concerning a problem in a particular area, then ask the question.

Ninety-nine per cent of the time *someone* will have the answer if you take the time and trouble to enquire. You will note that in several of my surveys, I made contact with various local authority officials. Sometimes they are unable to help, but they may know someone who does know the answer to a problem.

If in doubt, ask.

15 Using a moisture meter

See Fig. 15.1

GENERALLY

As you will have gathered from preceding chapters, the moisture meter is an essential tool for a surveyor to carry during a survey. This is because, whilst timbers might not be damp to the touch, they might well have enough water within them to encourage dry rot to germinate.

As wood is the most vulnerable part of the average house, standard meters are designed principally to test dampness in timber and are calibrated accordingly.

I would refer you to an excellent book on the subject, *Dampness in Buildings* by T. A. Oxley and E. G. Gobert (Butterworths), that covers the whole subject in great detail.

Oxley and Gobert divide typical moisture meters into two types, **conductance** and **capacitance**.

The type that I use is the 'conductance' type (i.e. Protimeter). All comments below should be treated as referring to this type of meter.

Moisture meters display readings either on a simple dial (as Fig. 15.1) or on a series of glowing lights (LEDs). The modern LED display has the advantage of being able to measure moisture levels from 6% up to 100%. Most meters are colour banded, green, yellow and red.

It is generally accepted that wood is 'safe' whilst its moisture level, as indicated by a moisture meter, is in the 'green zone'.

If a reading is obtained in the 'yellow zone', then there could be problems. Once readings are obtained in the 'red zone', then there is a very real danger of woodrot either being present or outbreaks occurring in the future, if the cause of the excessive dampness is not removed. (See 'Biological agencies', Chapter 12.)

HOW DOES A CONDUCTANCE MOISTURE METER WORK?

The 'conductance' type of moisture meter works by measuring the conductivity in the material tested. (The most common meters can be used to

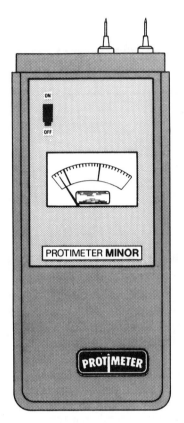

Fig. 15.1 Moisture meter.

measure conductivity in most materials but the readings are given in timber equivalent.)

The most common type of meter has two probes or spikes housed on the top. (There is usually a cap to cover the spikes when not in use.) Moisture meters also usually have another set of probes on the end of a cable for use in tight corners. These sharp spikes are a bit like 'Dracula fangs'. The steel 'fangs' are about half an inch long and are pushed into the material to be tested until firmly embedded. It is also possible to obtain substitute probes such as deep wall probes. These are very long probes which are insulated most of the way up the shaft to prevent surface readings. As the name suggests, they are used to take readings inside walls (after a hole has been drilled) so that the dampness readings can be checked in the centre of a wall.

As I have already said, the meter measures conductivity. (The fact that in the majority of cases the presence of water is indicated by high conductivity is beside the point.) It is important to remember that the meter is in reality similar to a voltammeter that you have in a car or battery charger, in that it is reading out the flow of current between the two points of the fangs. As indicated in earlier chapters, materials such as aluminium foil can create the impression that there is a high moisture content in the wall. If in doubt . . . Make several checks . . . If possible ease a bit of wall paper back to see if there is any other reason for a high reading.

WHAT TYPE OF MOISTURE METER?

What type of meter should be used during a Building Survey? As far as I am concerned, portability wins over bulk. The larger models might have greater accuracy but during a building survey you have enough to carry around and it is unlikely that the sorts of defect that are being sought out will be measured in very small percentages. It is also unlikely that destructive testing (i.e. drilling holes in walls etc.) will be allowed on the type of survey that I am describing. So why carry a very heavy model that has to be hung around the neck?

WHY USE A MOISTURE METER?

As I have said before, human senses are incapable of detecting the sorts of variation in moisture levels that are so essential when tracking down dampness. It is unlikely that you could determine by touch or sight a few degrees excess moisture, and it is those few degrees that could indicate that the timber is in danger.

HOW THE METER SHOULD BE USED

Most meters are designed primarily for use in timber and are calibrated in terms of actual moisture content based on the dry weight of timber. The reason for this is that timber is one of the softest parts of a dwelling and the one most liable to degrade and decay if neglected. However, a meter can be used to detect dampness elsewhere. For instance, if you are in a room and there is a circular stain on an external wall and it appears that the cavity is being breached at a specific point by, say, a cavity wall tie that has a 'snot' of mortar on it, you could check your theory by taking a series of readings around the stain. One would anticipate the moisture to slowly dry out the further away from the source. One would also anticipate that the spread would be uniform. In other words, the meter can be used as a check on your own senses and knowledge.

Faulty plumbing will also behave in a similar way if from an individual point source. The important thing to remember is that the surveyor obviously has to interpret the results of the readings.

RISING DAMP

As indicated in earlier chapters, rising damp is caused by the absence of, or a breakdown of, the damp proof course (DPC). Sometimes an adequate DPC can be bridged, if, for instance, a flower-bed were raised over the damp proof course level. If the DPC has failed, it is likely that there will be a damp zone above the skirting. There is a limit to the height that damp will rise under capillary attraction. Unless there are very special circumstances, it is unlikely that rising damp will climb more than 500–900 mm above ground floor level. The evidence of a 'tide mark' around ground floor rooms and high moisture readings usually indicate rising damp.

PENETRATING DAMP/DEFECTIVE RAINWATER GOODS

As indicated in earlier chapters, bad pointing on solid walls and/or feature projections externally can allow water to enter a building. Look for stains on wallpaper. Any suspect locations should be tested by inserting the prongs into the affected area.

On upper floors badly wetted walls could also indicate leaking gutters. Leaks on old parapet gutters are particularly likely.

RESIDUAL SALTS

When walls are dampened for a long time, hygroscopic salts (water absorbing) can still be within the plaster. If a wall is measured with a meter and it has salts within it, the readings are likely to be far higher than normal.

It is possible that a wall could be dry but the meter says it is excessively damp. However this is not necessarily a bad point as hygroscopic salts have the capacity to redampen the walls by absorbing water from the air. If these salts are present, walls that are re-decorated can quickly become discoloured again.

A positive reading therefore indicates that there is a condition that needs further investigation.

CONDENSATION (GENERALLY)

In previous chapters, I have intimated that condensation is a modern problem. Better insulation and fewer draughts mean that moisture levels can build up in the air and when the temperature drops in one area (e.g. on windows on a cold night) moisture condenses out (i.e. the dew point is reached). In bad cases, moisture condenses on walls and can seem like rising damp.

Where rising damp is the cause of wetness, skirtings usually give high readings. Where condensation is occurring the skirtings usually do not give high readings but the walls do. With condensation, mould growth and obvious causes of condensation are usually visible (e.g. tumble dryers, paraffin heaters, drying clothes, etc.), whereas with rising damp they are not.

CONDENSATION IN ROOF SPACES

As mentioned earlier in the book, condensation in roof spaces is becoming a common problem. Hot damp air rises in our highly insulated modern homes, escapes into a cold roof space and, being unable to escape any further (unless adequately ventilated in accordance with modern Buildings Regulations), it condenses in the loft, dampening timber and making the woodwork liable to attack. The moisture meter can provide information concerning the state of the roof timbers. But the surveyor must look for other signs to prove the readings. Look at the surfaces of the fibreglass quilt on the loft floor. If there are 'drip marks' beneath laps in the sarking felt, or actual drops of moisture on the felt or green/black mould visible in addition to high general moisture readings, then condensation is the most likely cause.

THE MOST COMMON USE FOR A METER

The most common use for the meter has to be on the skirting boards at ground level. (The skirting boards are the ones that run around the walls at floor level.) High readings on the skirtings and ground floor walls would *suggest* rising damp. I say 'suggest' quite deliberately. A high reading may have another cause.

Remember a moisture meter reads conductivity, and although moisture will undoubtedly make materials more conductive, so can other materials.

16 Pro-forma reports and RICS report outlines

RICS HOUSE BUYERS REPORT AND VALUATION

The RICS House Buyers Report and Valuation (HBRV) is a fairly recent development being introduced around 1981. It is *not* a structural survey because the standard wording on page one says so. It is held out as being an Inspection Report that provides a basic information on the general state of the property. It also informs the reader that the report will not detail defects of no structural significance or of a minor nature.

However, the general public ask surveyors to carry out an inspection for exactly the same reasons and therefore it would be inappropriate not to mention the existence of the HBRV.

SO WHAT IS THE DIFFERENCE BETWEEN A BUILDING SURVEY AND AN HBRV?

The HBRV basically comprises six standard sheets that have 35 sections. The surveyor fills in answers to 34 sections. The thirty-fifth section has a preprinted list of limitations included.

The first four questions relate to the client's name, address of the property, date of inspection and the weather conditions.

Questions 5 and 6 cover details on Tenure and Rateable Value.

From then on the report 'boxes' cover the property description and major elements of the structure.

One of the problems with the pro-forma report is that, if the surveyor wishes to provide more information than the box allows, then the client gets a report with 'see attached' in most of the boxes and an addendum is then attached to the report.

REPORT OUTLINES

In 1989 the Royal Institution of Chartered Surveyors encouraged a minimum standard of report content by the approval of Report Outlines. These

Outlines provide a library of typical descriptions for surveyors to adapt and use but do not profess to be able to teach people how to carry out surveys.

The Outlines are produced by Surveyors Software Ltd. (For details see Chapter 18, p. 167.)

When Report Outlines are used in practice, it is a requirement that a copyright acknowledgement is included in the reports produced and that a contract number specific to the Surveying Office concerned is quoted.

ADVANTAGES OF THE REPORT OUTLINES

The advantages are twofold:

(a) The surveyor or practice knows that he/she is conforming to an acknowledged standard of excellence if he/she uses Report Outlines as a basis for their documentation.
(b) The practice concerned can offer their client the advantages of the Building Defects Insurance. This insurance is not intended to protect the surveyor against claims of negligence. It is an additional cover should the client discover hidden defects that could not be reasonably expected to be found during a survey.

NB Report Outlines are copyright and can only be used under licence after paying the appropriate annual fee. None of the reports included in Appendices A–F are structured to comply with Report Outlines because they were written prior to the outlines being introduced. Any similarity of wording of any clauses included in the appendices is purely coincidental.

17 Using computer power

If you consider yourself 'computer literate', then this chapter may well be one that you might skip.

A computer rep. who I once met defined surveying as a 'word processing function'. At the time, it was something of a sales pitch, but I doubt if nowadays there are many surveyors' offices (of any type) that aren't harnessing the power of the computer in one way or another.

When considering Domestic Surveys, there are probably two main uses, which are:

(a) Word processing
(b) Structural calculations.

WORD PROCESSING

An old boss of mine had a motto: 'Never pioneer'. In other words, if someone else in the practice had carried out work of a similar nature, then the logical thing to do was to use the previous work as an example. This is one of the chief advantages of a word processing system.

For those who are not familiar with computers and word processors, it is difficult to explain the great assistance that these machines can be. However, I will try. The first question is what is a word processor?

The word processor

Strictly speaking, a word processor is a program that runs in a computer. In other words, the actual machine (the hardware) is a computer; the item that gives the computer its ability to deal with the input of text is a program (the software).

Instead of a document being typed onto paper, it is stored in the memory of the computer. When the document is complete it can then be turned into 'hard copy' (i.e. printed out from the computer).

Sometimes manufacturers muddy the waters slightly by describing dedicated computers as word processors.

How can a word processor assist report writing?

It was only a comparatively short time ago when computers in offices were the exception rather than the rule. Until fairly recently, some surveyors were highly sceptical as to whether or not the machines would really help their practices. Now few could do without them.

I can recall when everything was typed and re-typed (except for standard documents which could be photocopied). In many cases, the speed at which contract documents, reports and the like could be issued was largely dictated by the quality and/or speed of the office typist. The electric and later electronic typewriter speeded up the process but now we are in the era of the laser printer. The speed of reproduction is no longer measured in words per minute, or characters per second, but in pages per minute.

Before word processors, mistakes either had to be altered with correction fluid, re-typed or sent out in a less than perfect condition.

The word processor has the ability to use the computer's memory to store the typing. If some of the report is missed out or if more information needs inserting, the additional information can be inserted electronically, within the computer's 'brain'.

Reloading old text

The business computer has the ability to store old reports or a library of typical descriptions either on floppy discs or on a hard disc. These are in reality very sophisticated tape recorders. (Did I hear someone who knows something about computers suddenly breathe in sharply? Okay, so floppy discs and hard discs have random access facilities . . . but so what! I am trying to explain to those who have never used a computer. They record data for future use.)

This means that the office typist no longer has to start from the very beginning every time that a new report comes in. The library or an old report can be loaded into the computer's memory and adapted.

The word processor saves typists' time and the surveyor has the advantage that he is not 'pioneering'.

The dangers of the word processor

In computer circles there are two expressions that seem appropriate and these are 'drop-out' or 'drop-in'. Once again, the purists will probably shout!

'Drop-out' is when vital information is lost from a record; and 'drop-in' is when something that shouldn't be there suddenly appears. There can be many reasons for this such as a 'power spike' or a disc fault but there is also the human cause, especially where reports are concerned.

Say you have a perfect 'library' of descriptions and someone who doesn't know what they are doing omits a vital clause. From that point onwards, it is possible that every report could be deficient. The clause has 'dropped out'. It no longer exists, and until someone notices, it will remain so.

It is also possible that unwanted clauses could be included if, say, two files are 'merged'. Imagine the embarrassment if the description of a cellar were included in a report for a house that hasn't got one.

It couldn't happen? I bet it has!

The moral

Word processors are a very useful tool but you should nevertheless be on your guard and read over the report very carefully to prevent mistakes getting through.

STRUCTURAL CALCULATIONS

As in the case of the Ninetails Avenue report (p. 189), there are times when there could be a need for structural members such as timber beams and RSJs to be checked for strength.

Under normal circumstances, the Building Regulations will provide enough information with regards to joist, rafter and purlin sizes, etc. (see tables B3 to B28) but there are instances where the Building Regulations do not cover the item in question. Steel beam sizing is one such item.

On large spans and very heavy loadings checking this type of detail is undoubtedly a task for a qualified structural engineer but for smaller spans, once again, the computer comes to our aid. If the survey creates a need for structural calculations to ensure that no part of the structure is being overstressed, then there are a variety of programs available for producing these calculations. As I am familiar with the system I intend to use the program *Pocket Engineer* as an example (see Chapter 18, p. 168, for details).

This program deals with calculations for steel and timber beams used to support part of the building above an opening or an overhang. There is a restricted range of beam sizes (Fig. 17.1), but as I have said, it would be unwise for surveyors to trespass too far into engineering.

The program offers a choice of various loading situations (Fig. 17.1). All that is required is for the user to choose the most suitable situation (i.e. cantilever, point loads, distributed loads and/or free and tied supports).

The likely loading on the beam is then calculated using a standard table which can be edited to suit the calculation you require. (Standard data is

162 Using computer power

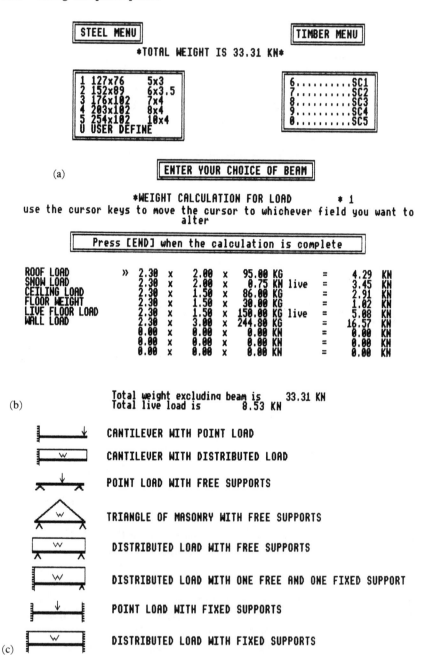

| STEEL MENU | | TIMBER MENU |

TOTAL WEIGHT IS 33.31 KN

```
1 127x76    5x3
2 152x89    6x3.5
3 176x102   7x4
4 203x102   8x4
5 254x102   10x4
U USER DEFINE
```

```
6..........SC1
7..........SC2
8..........SC3
9..........SC4
0..........SC5
```

(a)   ENTER YOUR CHOICE OF BEAM

*WEIGHT CALCULATION FOR LOAD   * 1

use the cursor keys to move the cursor to whichever field you want to alter

Press [END] when the calculation is complete

```
ROOF LOAD       »  2.30 x  2.00 x   95.00 KG      =   4.29 KN
SNOW LOAD          2.30 x  2.00 x    0.75 KN live =   3.45 KN
CEILING LOAD       2.30 x  1.50 x   86.00 KG      =   2.91 KN
FLOOR WEIGHT       2.30 x  1.50 x   30.00 KG      =   1.02 KN
LIVE FLOOR LOAD    2.30 x  1.50 x  150.00 KG live =   5.08 KN
WALL LOAD          2.30 x  3.00 x  244.80 KG      =  16.57 KN
                   0.00 x  0.00 x    0.00 KN      =   0.00 KN
                   0.00 x  0.00 x    0.00 KN      =   0.00 KN
                   0.00 x  0.00 x    0.00 KN      =   0.00 KN
```

(b)   Total weight excluding beam is   33.31 KN
           Total live load is   8.53 KN

CANTILEVER WITH POINT LOAD

CANTILEVER WITH DISTRIBUTED LOAD

POINT LOAD WITH FREE SUPPORTS

TRIANGLE OF MASONRY WITH FREE SUPPORTS

DISTRIBUTED LOAD WITH FREE SUPPORTS

DISTRIBUTED LOAD WITH ONE FREE AND ONE FIXED SUPPORT

POINT LOAD WITH FIXED SUPPORTS

(c)   DISTRIBUTED LOAD WITH FIXED SUPPORTS

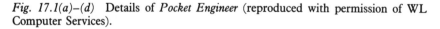

Fig. 17.1(a)–(d) Details of *Pocket Engineer* (reproduced with permission of WL Computer Services).

```
BEAM MOMENT CALCULATION FOR  Demonstration Purposes
BEAM LENGTH IS  2.00 metres
Timber grade is          SC4
Beam width is    100.00 mm
Depth is         297.97 mm
```

```
LOADING CALCULATION FOR LOAD     1.00

    ROOF LOAD        2.30 x    2.00 x   95.00 KG        =    4.29 KN
    SNOW LOAD      - 2.30 x    2.00 x    0.75 KN live   =    3.45 KN
    CEILING LOAD     2.30 x    1.50 x   86.00 KG        =    2.91 KN
    FLOOR WEIGHT     2.30 x    1.50 x   30.00 KG        =    1.02 KN
    LIVE FLOOR LOAD  2.30 x    1.50 x  150.00 KG live   =    5.08 KN
    WALL LOAD        2.30 x    3.00 x  244.80 KG        =   16.57 KN
                     0.00 x    0.00 x    0.00 KN        =    0.00 KN
                     0.00 x    0.00 x    0.00 KN        =    0.00 KN
                     0.00 x    0.00 x    0.00 KN        =    0.00 KN
```

```
Total weight excluding beam is     33.31 KN
Total live load is  8526.67 N
Loading stresses are:-
```
The depth required to resist bending is found by equating the Bending Moment (BM)
with the Moment of Resistance (M.R.) where
    M.R. = (breadth x depth x depth)/6
The depth required to resist shearing is found by equating the shear force with
the allowed shear stress given by
    Allowed shear = shear stress x breadth x depth
The depth required to resist excessive deflection is found by ensuring that the
actual deflection is less than one in 360 of the length where the moment of inertia
is given by $I = \dfrac{breadth \times depth3}{12}$

```
Load     1.00
Load is              33294.87  N
Live load is          8526.67  N
```
Maximum bending moment is $\dfrac{\text{Load x Length}}{6.00}$ = 11098291.36 Nmm

Shear force is $\dfrac{\text{Load}}{2.00}$ = 16647.44 N

Minimum depth to prevent excessive deflection is   =   154.96 mm
    Where deflection is given by $\dfrac{\text{load x length3}}{60.00 \times \text{Elastic Modulus x Moment of Inertia}}$

```
Total bending moment is                            11098291.36 Nmm
Minimum depth required to resist bending is        297.97 mm
Total Shearing stress is                           16647.44 N
Minimum depth required to resist shearing is       234.47 mm
Minimum depth required to prevent excessive deflection is  154.96        mm
```

```
The beam width is 100.00mm
```
(d) The length of the overlap of the beam shall be 469.15mm

Fig. 17.1   Continued

offered within the program. There is no need to remember or enter weights of standard materials, but the weights and forces shown can be amended if the user so chooses.)

A typical case is shown in Fig. 17.1. The result can be printed out for permanent record purposes (Fig. 17.1). The advantage of using a package such as this is the ease with which the figures can be entered and adjusted to suit requirements, providing for you the standard loads and stresses for each beam without having to look them up each time.

Software such as this can provide the necessary calculations within minutes as opposed to working for perhaps an hour by hand. For those interested, further information can be obtained from WL Computer Services (Tel. 051 426 7400).

18 Conclusion

WHAT MAKES A GOOD SURVEYOR?

Obviously, you have to be interested in what you are doing, be observant, have a good knowledge of building construction and be analytically minded.

The first characteristic is not of your own choosing. You either like buildings/the building industry/and builders or you don't. If you hate everything about the industry (which I would find surprising since you are reading this book), then forget it. It is not for you. Hopefully, you are interested and this book has aided in your progress.

The next ingredient for the mix is being observant at all times. I say 'at all times' quite deliberately. Very few people spend all their time carrying out surveys, but for those involved in the building industry in the widest sense, the observant person can pick up large amounts of knowledge during a working day. These nuggets of information can then possibly be applied to the survey work. But be sure, however, that you analyse what you see. It is as easy to pick up bad habits as good ones. The non-analytical mind can accept bad site practices, instead of condemning them.

Knowledge (other than that gained above) can only be achieved by study, but once again, a real interest in what you are doing makes study far easier.

Finally, when carrying out a building survey, although you are concerned about how the building should have been constructed, more important is how the building *has* indeed been constructed. The older the property, the less likely that all aspects of construction will be perfect. In any case, poor building will always be with us. The person carrying out a survey is basically a detective; and although some might disagree, in my opinion, there is little difference between the surveyor checking out a house before someone buys it and the AA mechanic who gives a car an inspection for latent defects. The only problem for the surveyor is that he/she can't run a house over a pit and tinker around with the underside. Like everything about building, it has to be done 'on site' and the hidden bits remain hidden.

Don't forget, though, that you are not entirely alone. As I have intimated previously, the Building Control Department at your local council can be of great assistance if courteously approached. They know of problem 'pockets' in their area and will probably provide some 'background' if they possibly can. British Coal has records that cover the country. If it is applicable, use their services. I also ensure that I keep a copy of each and every Mining Report that comes into the office. In theory, as long as they don't go out of date, you could have detailed underground records that cover very large areas.

Then there are you own observations in an area. If you have driven around a particular area and have noted that large numbers of properties seem affected by settlement, then use this knowledge when formulating your report.

Don't forget your own records. If you have done a survey in one location and a defect has been found, the likelihood is, especially on a large estate built by the same contractor, that similar problems might be experienced by similar houses on the estate in question.

If one property has a sulphate attack on the floor, then the house next door is likely to be subject to a similar problem. After all, the men who worked on the two houses were probably the same and if not properly trained or supervised, will undoubtedly have taken the same 'short cuts' on adjacent properties. Then there are areas with a particular problem, as highlighted in the Ninetails Avenue report (p. 189).

Building Surveys are part science and part art. The surveyor must develop both facets. The public expect the surveyor to be the expert, who, by his/her training, should be able to work out if and why a property has defects and the likely remedies.

To do the job properly, protective clothing should be worn over your good clothes. Someone who is only concerned with maintaining sartorial elegance is unlikely to be the sort of person to do a good survey!

I would caution that this book only covers domestic situations. I do not envisage my readers embarking upon surveys of multi-storey buildings, industrial and/or high rise structures that are of non-traditional construction. I would further add that before even thinking of embarking on survey work in a professional capacity you should consider adequate professional indemnity insurance. Recent RICS statistics show that Domestic Surveys are a 'high risk' area. Over 40% of all Professional Indemnity claims against surveyors are in connection with Residential Surveys.

Finally, when carrying out Building Surveys the reader should be on his guard for items that just don't 'feel' right. If they don't, then there probably is something wrong. It is a case of your inner self trying to warn you. Don't try to walk away from the problem. Worry it until you are satisfied that you have all the answers. *If in doubt . . . check . . . and then double check.* It is better to crawl through a dirty roof void twice and be right than skimp and regret it at leisure.

THE RIGHT TOOLS FOR THE JOB

In order to carry out a proper survey you will need to have the right tools for the job. There may be others that you would add to the list but here are the basic essentials:

(a) Powerful lamp and long lead/torch.
(b) Ladder not less than 3 m in length.
(c) Mirror (one on a 'stalk', good for looking at the backs of pipes).
(d) Sharp probe (such as a penknife or sharpened screwdriver).
(e) Opera glasses (for checking on roofs and the like).
(f) Moisture meter.
(g) Jemmy (for lifting floorboards).
(h) Crowbar.
(i) Screwdrivers.
(j) Manhole keys.
(k) Long tape measure.
(l) Notebook and/or tape recorder.
(m) Plumb line.
(n) Long spirit level.
(o) Hammer.
(p) Wide bladed chisel.

NEARLY THE FINAL WORD

No book can condense a subject like Domestic Surveys into one volume. For your assistance, I have listed some other useful contacts, books, leaflets and names and telephone numbers:

Recommended list of contacts/books/leaflets/for further information/reading

Professional bodies
The Chartered Institute of Building
Tel. 0344 23355

The Institute of Building have a large library of technical books which can be borrowed by members of the Institute.

Royal Institution of Chartered Surveyors
Tel. 071 222 7000

There are a wide number of leaflets available from Surveyors Publications which cover the terms and conditions for Surveys and House Buyers Reports.

Documents/books
RICS Report Outlines. Surveyors Software Ltd. Tel. 0603 622466

The Building Regulations 1985 and 1990/92 Amendments. HMSO, London.

NHBC Standards. The National House-Builders Council, 58 Portland Place, London W1N 4BU.

Dampness in Buildings. T.A. Oxley and E.G. Gobert. Butterworths, London.

Surveying Buildings, Malcolm Hollis. RICS, see above.

Green Book Design Manual. Pilkington Insulation. Tel. 0744 24022

Ventilation Products. Willan/Glidevale. Tel. 061 973 1234

Roofing Products. Willan/Glidevale. Tel. 061 973 1234

Building Reference Book. Northwood Books, London.
Building Research Establishment Digests. Tel. 0923 664444
Digest 329. Installing Wall Ties in Existing Walls.
Digest 299. Dry Rot its Recognition and Control.

Computer software
WL Computer Services. Tel. 051 426 7400
Several useful building orientated computer programs are available that run on IBM and Acorn systems.

Surveying instruments
Protimeter PLC. Tel. 06284 72722

THE FINAL WORD

I hope that this book will be of assistance to those of you who decide to involve yourselves with Domestic Building Surveys. It is my hope that I have covered as much ground as possible, but as I have said above, it is virtually impossible to cover every eventuality. There are always new problems that arise and you must try to keep abreast of these developments.

APPENDICES: Domestic Building Surveys

Appendix A: 'The Wetlands' report

For

'The Wetlands'
75 Church View,
Tipham,
Tipfordshire

For

Mr C. Vernon,
Lady House,
Threadbare Street,
London

Building Services (Technical)
Old Ferry Offices,
The Square,
Tipham,
Tipfordshire

March, 19—

NOTE
*THIS REPORT HAS BEEN INCLUDED TO ILLUSTRATE THAT A
BUILDING SURVEYOR DOES NOT NECESSARILY ACT ALONE.
IN THIS CASE THE REPORT WAS A COMBINED EFFORT
REQUIRING INPUT FROM AN ENGINEER, AN ELECTRICIAN
AND A PLUMBER AS WELL AS THE SURVEYOR.*

Although the following report is based upon a genuine survey, the names of
any persons, companies, properties and the locations have been deliberately
changed. Building Services (Technical) is a registered business name (Reg.
no. 2175590).

The Coal Board report, engineer's report, the electrician's and plumber's
reports, and the various sketches and documents referred to in the text of
the main report, have not been reproduced here.

References to sketches in the report have been dashed out in order that there is no confusion with those referred to in the rest of the book. The sketch layout of 'The Wetlands' is included at the beginning of Chapter 4.

The report has not been revised to allow for recent changes in the Building Regulations. Note in particular that recommended thicknesses of insulation have been changed since the original report was written.

BUILDING REPORT ON 'THE WETLANDS', 75 CHURCH VIEW, TIPHAM, TIPFORDSHIRE

INSTRUCTIONS, LIMITATIONS AND GENERAL PREAMBLE

We report that we carried out a survey at the above property on — March, 19— as per our letter of confirmation dated — March, 19— and our standard conditions of engagement (see appendix).

This report shall be for the private and confidential use of our client for whom the report is undertaken and must not be reproduced in whole or in part or relied upon by a third party.

We will not be responsible to third parties who obtain, by any means, a copy of this report and act upon information contained therein.

As you are aware, prior to carrying out our main survey, we visited the property with our structural engineer to examine the settlement cracking on the front and rear elevations and the tie bar on the front elevation. His findings have been bound into the rear of this report (see appendix).

Whilst every care has been taken in completing our survey and report, the investigations have been non-destructive in nature and therefore we are unable to report upon matters which were concealed at the time of the inspection and the following assumptions have had to be made:

(a) That high alumina cement (HAC), concrete or calcium chloride additive or other deleterious materials or techniques were not used when the original property or any subsequent extensions were built.
(b) That any wall ties that exist are not perished.
(c) That your solicitor's or legal adviser's search will prove that the property is not subject to any unusual covenants, Local Authority restrictions, or onerous restrticions imposed by others, encumbrances or outgoings and that good title can be shown.
(d) That future inspections of any hidden parts of the structure which have not been inspected during this survey will not reveal material defects, or cause the surveyor to alter materially the indicative costings shown at the end of this report (if applicable).

We have not inspected woodwork or other parts of the structure which are not uncovered, unexposed or were inaccessible. Neither have we removed insulation from between or laid over ceiling joists to inspect the hidden surfaces of the timber because such an undertaking involves extensive builders' work which does not form part of our conditions of engagement. We are therefore unable to report that the property is free from defect. However, we have done our best to draw conclusions about the construction from surface evidence visible at the time of our inspection. (*NB* Later in the report you will note that we found outbreaks of woodworm and woodrot in the basement and roof space.)

Whilst we have tried to amplify our descriptions when technical terms have been used in this report, we would also direct you to our standard cross-section through a typical house and sketches bound into the rear of the report.

The plumber's report, electrician's report and Coal Board Report which were carried out in conjunction with our inspection are also bound into the rear of this document.

As promised, we have included indicative costings to cover likely remedial works. However, we would stress that these are only intended to be an approximate guide and not an exact cost. They are given so that you understand the likely implications of the purchase in approximate monetary terms. As the costings are based upon average building prices, assumed specifications and assumed areas, some of the estimates may be high and some may be low. It is essential therefore that competitive estimates are obtained from several local builders prior to instructing work to be carried out. Our costings should only be treated as a guide, current at the time of preparing this report with no allowance made for value added tax.

References to left and right are made facing the house, standing in Church View. The garage, for example, is on the right of the house.

THE PROPERTY GENERALLY

The property is a large semi-detached house (two storey plus cellar/ basement). It contains two living rooms, a dining room and a kitchen on the ground floor and five bedrooms and a bathroom on the first floor. The cellar/basement is divided into three rooms which at the moment have no specific function.

Based upon information sheets that we sent to the vendor prior to our visit and photocopies of old documents provided by the vendors, we have good reason to believe that the house is approximately 150 years old. (See document bound into the rear of the report.)

The rear garden of the house backs onto a small lake in Greenman Park. As you will note, the document suggests that the house and the park were originally part of the Greenman Estate.

You were concerned about the possibility of flooding in this area and specifically requested that we made enquiries on this point.

We believe that the living accommodation (i.e. ground and first floor) is unlikely to by affected by such an eventuality because the Local Authority informed us that improved surface water drainage was installed as part of the South Dean Underpass project. However, there does appear to be evidence of flooding and dampness in the basement floor and walls. We have enlarged upon this matter later on in the report.

The main roof of the house has been covered with hessian and bitumen according to the current owner. (From now on the terms 'treated' or 'treatment' will be used when describing the hessian and bitumen covering.)

During our site inspection, we were shown invoices and correspondence from the firm that carried out the roof treatment indicating that the work was put in hand just over two years ago. We understand that in order for guarantees to remain effective that the treatment be renewed every —— years. We have enlarged upon this matter later on in the report.

As you are familiar with the property and have the estate agent's details to hand, we have not provided superfluous information such as room sizes, the exact location of the property in relation to schools in the area, bus services, etc. However, should there be any items that you wish us to clarify further we will endeavour to do so.

Our overall impression of the property was that it was in need of a great deal of maintenance.

THE MAIN STRUCTURE

The main pitched roof

Long ladders and roof ladders have not been used for inspection.

Where practicable the interior of the main roofs have been examined but the inspection of the roof covering has been made from ground level only, using binoculars.

It was not possible to inspect the interior of the lean-to roof because of restricted access.

One ridge tile appeared to be missing on the main roof but we believe that this location may be 'treated' over. (See 'The property generally' for details.) However, it is advisable to have this examined and replaced by a competent roofer.

Internally, both main roofs are traditional rafter and purlin construction. The cement/lime render underneath the slates has to a large extent fallen off. Under normal circumstances, without the render (which is a secondary defence against water penetration), any missing slates would immediately let water into the roof. It was probably roof leaks that encouraged the present owner to have the roof 'treated'.

Except for the low level roof to the rear, very little of the original slating could be seen because the surfaces had been 'treated'. On the low level roof to the rear extension, slates were missing on the verges and need replacing (see sketch ——).

The main roof is in two distinct sections with a secret gutter in between. (See sketch —— for location.) Because two of the slopes are hidden from view from ground level, it is not possible to comment upon them. It is worth bearing in mind, however, that a leaking secret gutter can be a source of present or future trouble. There is evidence of a past leak in the front right-hand bedroom and on the landing.

From information sheets filled in by the vendor and returned to us prior to our visit, we understand that this defect was discovered some five years ago and was rectified within days of the first signs of leakage becoming apparent. A small outbreak of wet rot was discovered whilst tracing the leak and was also treated. We have bound into the rear of the report a copy invoice from Messrs Fixit Quick (Often Narrows) Ltd that would substantiate that the defects were rectified (see document). Please also note that the contractor's guarantee is still in force, has five more years to run and is transferable.

The small pipe projecting out of the roof on the rear elevation that you specifically enquired about would appear to be the expansion pipe for the central heating system. In more up-to-date systems this would discharge over a header tank inside the loft. The present arrangement could freeze up in very cold weather and it would be wise to have this arrangement altered, if you purchase the property. (Also see plumber's report enclosed.)

Access was gained to the rear main loft via a trapdoor in the bathroom. The current owners have constructed a secondary suspended ceiling below the original and this made access very difficult. Once within the main rear roof void, movement was impeded still further by a badly sited water tank.

We gained access into the front loft from the landing. Once again this trapdoor was difficult to use.

If you do purchase the propery, we would advise you to have the trapdoors widened by a joiner and roof ladders installed so that proper maintenance of the loft areas can be adequately carried out. It would also be advisable to have the water tank resited. The likely cost of this work would be £. . . . .

The ceiling of the front loft had been recently insulated with 100 mm thick quilt. This had been laid over an existing rockwool mat which was approximately 1–2 in (25–50 mm) thick.

Unfortunately whoever laid the insulation did so in such a way that the ceiling joists are now hidden from view. We considered it unsafe to try to cross the ceiling in the front loft area. This made a thorough examination of roof timbers in this area impossible. Luckily, in one section there were crawl boards and we went along these towards the right and inspected the rafters within easy reach of the crawl boards.

Subject to the limitations of the inspection in the front loft and in the 'lean-to', our comments on both areas are as follows.

We found signs of woodworm in some timbers inspected (by the rear bathroom access) and what appeared to be an outbreak of true dry rot

(*Serpula lacrymans*) (see sketch for location). As there were high readings given by the moisture meter (about 25%) in all areas, the risk of further outbreaks of dry rot being found is high, in our opinion.

It is generally agreed by most timber experts that once timber moisture exceeds 20% then the chances of a *Serpula lacrymans* (true dry rot) outbreak become possible. It is essential therefore that the outbreak found is treated by a competent firm of dry rot eradicators as soon as possible. They should be instructed to inspect the whole roof void for further areas of rot and not just the area indicated.

If this outbreak is left untreated, the roof timbers will be very quickly eaten away and will disintegrate within a very short space of time because dry rot can grow very quickly given the right conditions. (We have heard an expert mycologist express the opinion that it is possible for the fungus to grow by 18–25 ft (6–8.5 m) per annum when given optimum conditions in a controlled laboratory experiment.)

Dry rot, unlike other fungi found destroying the fabric of a building, has the special ability to pass through and over masonry or plaster and can infect parts of property some distance from the seat of the initial outbreak. However, dampness in itself is not usually enough to set off an attack because the fungus likes poor light, moisture levels in the timber of above 20%, bad ventilation and warm surroundings. The important thing to do, therefore, after treating the initial outbreak, is to make the conditions disagreeable for the fungus.

The high moisture readings would suggest that either the roof coverings were defective or that there was condensation in the roof void. As the roof has been 'treated' it is unlikely that the roof is leaking now. Therefore condensation is the most probable cause of the damp.

As previously mentioned dry rot fungus hates good ventilation and needs some moisture to germinate. We would strongly recommend the incorporation of several patent roof ventilator slates into the various roof slopes to release the build up of warm, damp air. These ventilators are relatively cheap to install and will let moisture vapour out of the roof without letting rainwater in. The increased circulation of air will reduce moisture levels in the wood, over a period of time, by evaporation. Good cross-ventilation will prevent the conditions that dry rot needs to thrive re-occurring. The likely cost of installing these ventilators would be £. . . . .

Chimneys (3 no.)

The chimneys were viewed through binoculars from ground level and seemed in need of repointing. The mortar bedding to the chimney pots also appears to require renewal. The rear right-hand chimney had a lean and will require rebuilding. This should be treated with urgency as the stack could easily fall in a high wind. There was damp in the cupboard in the rear left-hand bedroom (as viewed from the road) which would indicate that the flashings and soakers to this stack require replacing. (Flashings and soakers are purpose made pieces of lead or lead substitute that close the gap between the slates and the chimney.)

We understand that one chimney has already been removed by the present owner. The usual reason for the removal of redundant chimneys is to prevent damp entering the building. The flue to the redundant chimney does not appear to have been vented which is bad practice as not doing so can encourage dry rot outbreaks. Airbricks and plaster grills should be installed as per sketch . . . . . bound into the back of this report. The likely cost of remedial work to the chimneys will be £. . . . .

Gutters, downpipes, plumbing and drainage

The property did not appear to have its own access manhole (which is not unusual with a house of this age), so it was therefore not possible to decide whether a separate foul and surface water system served the dwelling. There was a metal plate adjacent to the garage which we lifted, believing it to be a manhole cover, and found that it did not appear to serve any purpose.

The gutters at the rear of the propery at high level and above the bay to the rear lounge have been replaced by a new PVC system. The downpipes however were the original cast iron pipes and we found a large split in the one adjacent to the small window to the rear lounge. Some work will be required on the downpipes and we would allow £. . . . . for this work.

At the front of the house, the wooden gutters and fascias have rotted and will need replacing. The costs of rectifying this work will probably be in the region of £. . . . .

The vent pipe on the right-hand side adjacent to the old tank is badly holed and has been bandaged. This needs replacing entirely. The likely cost will be £. . . . .

Main walls of house

No examination has been made of the foundations because this would require the excavation of inspection pits.

There was no evidence that the house had been built with a damp proof course (DPC). As you are probably aware, DPCs are incorporated into all modern properties in order to prevent rising damp and damage to the structure of the buildings. This omission is one that *must* be remedied.

We would recommend that an injection damp proof course be installed in the walls by an approved firm. (We can let you have a copy of our approved list should you so require.)

An injection DPC to the walls would cost approximately £. . . . . and the cost of replastering 1 m high with approved renovation plaster or a plaster approved by the injection company £. . . . . (We have not allowed for the cost of redecoration as we have assumed that you will be redecorating the whole house to your choice.)

We would particularly bring to your attention the necessity to replaster the ground floor walls for the first metre with special plaster. (Not normal lightweight plaster.) If this is not done, the water absorbing salts in the old plaster will inevitably damage the new decorations and give the incorrect impression that the walls are still suffering from rising damp.

The external walls of the house are constructed in a form of English garden wall bond (three courses of stretchers to one of headers) which would indicate that they were built of solid and not cavity construction. Also, the age of the property would make cavity construction very unlikely. From measurements taken, it would appear that the external walls are solid 9 in (225 mm) brick walls with 14 in (350 mm) walls to the basement area.

The reason why we stress the wall construction is that cavity walls are nowadays considered the norm under modern Building Control Regulations. The reason for this is because cavity walls are far less likely to suffer from penetrating dampness which was a common fault with solid construction.

At the base of the external walls there was a cement plinth at ground level. The cement plinth is an old-fashioned way of trying to prevent surface water from entering a building and tends to become ineffective very quickly once the render coat cracks or becomes detached from the wall. Water can then lodge behind the plinth and work its way through to the inside of the house.

We would also recommend that the loose render plinth externally be removed (particularly adjacent to the new bay window to the rear lounge) as it is now trapping water rather than keeping it out.

To help keep out penetrating water ingress we would also recommend that the entire house be repointed on all elevations. Two airbricks at high level on the rear elevation are damaged and require replacing.

The likely cost of the plinth, the air bricks and the repointing work is £. . . . .

Windows, doors and timberwork

The windows generally appear to be the original single glazed sash type window and all require mastic pointing at abutments with the brickwork. We have not tested the window sashes in all areas (they were burglar proofed or nailed shut), but we noted that two windows require re-cording and would suggest that most of the windows will require some maintenance work both re-cording and re-weighting and cutting and splicing in new wood where found rotten.

In particular we would report that the small window to the rear lounge has a 'bullet hole' in it and the frame is particularly rotten and there are minor settlement cracks over the arch. (See engineer's report.) The bay window to the rear lounge has recently been replaced and is in good condition but requires painting. The back door is 'Canadian style' and has a fly screen. We noted that there was wet rot in the feet of the door frame and new timber will need to be spliced in. The front door and frame gave high moisture readings but were sound. The glazed area below the front door step is letting water into the basement and we would suggest that the gap between the step and the glass be sealed with a mastic joint. The likely cost of remedial work to doors and windows will be in the region of £. . . . .

Sundry external woodwork

We examined fascia boards etc. on all elevations through binoculars and they seemed to be in a very bad state of repair. There would appear to be extensive wet rot and we would suggest that all this woodwork will require renewing. (This has been costed elsewhere.)

External painting

The cost of repainting the external woodwork would be in the region of £. . . . .

INTERIOR GENERALLY

Decoration

The property generally was in a reasonable state of decoration.

Floors

The cellar floor was either concrete or stone flagged.

The ground floor was partially solid (kitchen area) and partially suspended timber. The first floor was suspended timber construction. We were unable to remove floorboards anywhere, and we could only lift corners of carpets, so we are unable to comment upon the condition of the timber floors except to say that in the areas observed all seemed to be in order.

Ceilings

The ceilings were generally in reasonable condition except for the odd crack. On the landing and the front right-hand bedroom there was evidence of a roof leak. (This matter has been dealt with elsewhere.)

Electrical installation

From inspecting the system it would appear to be the old radial type. The up-to-date electrical regulations recommend a ring main which allows for expansion. The wires would seem to be perished and we would recommend a complete rewire. The likely cost would be £. . . . . (See also electricians report).

The points noted were as follows:

	Single sockets	*Light points*
Ground floor		
Kitchen	2	2
WC, cooking area	—	2
Hall	—	2
Rear lounge	4	1
Front room	3	1
First floor		
Landing/staircases	—	1
Front left bedroom (box)	1	1
Front right-hand bedroom	1	1
Rear right-hand bedroom	1	1
Side bedroom	1	1
Rear left-hand (oriel window)	2	1
Bathroom	—	1
		(Heat/light)
Cellar	None noted	
Generally		1 per room (except rear right box room)

Water services/heating installation

These systems seemed in good working order. We turned on all taps and they appeared to work. There were radiators in all habitable rooms (none noted in side bedroom, small left-hand front bedroom, downstairs WC or kitchen). The kitchen had an Aga which provided quite a bit of background heat. (See also plumber's report.)

FIRST FLOOR

Bathroom

The WC seemed a little faulty when flushed and we would suggest that the cistern needs overhauling by a plumber. The wash basin was cracked and the wall plaster was uneven. A replacement sink would not seem necessary unless you particularly wish to renew same.

Landing/airing cupboard

The stained ceiling has been dealt with elsewhere. The water tank in the cupboard on the landing was badly corroded inside and requires replacing. The doors to the cupboard were stiff and require easing. The likely cost of both operations would be £. . . . .

There is a slight fall to the floor which would suggest that there has been differential settlement in the past. This fact was reinforced because the staircase has moved slightly away from the wall. (See engineer's report.)

Small front bedroom

No further comments.

Right-hand front bedroom

Comments on stained ceiling elsewhere. There appeared to be a section of defective wall plaster near the window. The cost of replacing this would be approximately £. . . . . No further comment.

Rear bedroom/living room

Sash cord broken. No further comment.

Right middle bedroom

No further comment.

Rear left bedroom (with oriel window)

Dampness to chimney stack in cupboard reported previously. No further comment.

GROUND FLOOR

Rear lounge

The tiles to the hearth were loose. No further comment.

Front room

There is evidence of a leak to the bay window. When questioned, the present owner informed us that this had been repaired some years previously and an invoice was produced which indicated that this was so.

Kitchen

The kitchen floor was red quarry tile and seemed to have a slight slope on it which, once again, supports the belief that differential settlement has occurred in the past. The skirtings were not timber and appeared to be cement and sand. The electrical system was overloaded with appliances. There is a ceiling mounted clothes airer near to the Aga. The kitchen units are old and require replacement. The likely cost would be £. . . . . if Messrs . . . . . units were installed. (See comments below on cost of new floor).

Downstairs WC/cooking area

The ceiling was badly cracked and might require plating with a layer of plaster board and skimming. The likely cost of this work would be £. . . . . The feet of the door frames had high moisture readings which would suggest that the red quarry tiles are also under this area. The typical construction for kitchen floors in the past was to lay quarry tiles on a thin screed direct on compacted hardcore. There would have been no damp proof membrane provided and as a consequence, this type of floor sweats when lino or other finishes are laid on top. The only way of totally waterproofing this type of floor is to dig it up and relay it using modern construction. The likely cost of relaying the kitchen/WC/cooking area floor would be £. . . . .

The toilet has been put in recently (according to the present owner). Both the toilet and Belfast sink seemed to work. The window to the toilet has two layers of glass. The obscured glass is tacked in behind the original pane and children could easily cut their hands on the exposed edges.

Hall

No further comment.

CELLAR/BASEMENT

The cellar comprises several rooms (see sketch for location) and judging by the construction was probably used as working quarters for the servants when the house was originally constructed. The cellar floor generally comprises stone flagging. There were two sumps in the basement area (pockets in the floor that collect water) but the pumps were not working (the electrician has been able to confirm this). From these sumps a series of channels radiated outwards and we came to the conclusion that the basement has always been subject to flooding and that when the water level rose, the electric pumps cut in and lifted the excess water into surface drains. Judging by marks on the whitewashed walls, we believe that in wet weather, the basement regularly fills with water to a height of approximately 200 mm (8 in) above basement floor level. It is essential that the pumps and drainage system are renewed. The likely cost of this would be £. . . . .

The walls are whitewashed brickwork and the ceilings are the exposed timbers of the floor above. To the rear, the rooms were originally ventilated to the open air, by way of light wells. These have been blocked off at ground level by the present owners who informed us that they had done so to prevent damp getting into the basement. The old wells still remain and, in our opinion, are a danger to children because they will no doubt be used as hiding places. If the window sashes were to drop whilst a child was playing in the void beyond, it is possible that he/she could remain trapped for some considerable time.

An outbreak of dry rot was noted in one of the light wells which gives cause for concern. (See location plan.) It is obvious that the conditions in the cellar area generally are ideal for further outbreaks.

The room under the front living room is ventilated by a large pipe which leads up to the outside air. We believe that judging by its position it is connected to the defective vent pipe by the oil tank at the side of the house.

There are signs that rainwater is running into the cellar down this pipe. It is essential that some sort of waterproof ventilating cap be placed over the pipe to prevent further water entering whilst maintaining the essential ventilation. There are also smaller ventilators in the walls beneath the floor joists but these seemed to be blocked up deliberately. These require cleaning out so that the fresh-air circulation is improved. If rainwater enters the cellar/basement through these airbricks, then the external ground level should be lowered slightly.

There is evidence of a small woodworm outbreak (see sketch for location) which was probably introduced by a log (which has since been removed) and there is also evidence of woodrot in one of the floor timbers closest to the damp external wall. The ledged and braced door in the rear room also had holes in it but owing to the formation we concluded that a dartboard had been hung there at some time in the past. The left-hand rear room had rotted timber above the window. (This is the window adjacent to the dry rot outbreak.)

The whole cellar had areas where the bricks had crumbled (spalled) and we would advise that all the spalled bricks be cut out and replaced by new ones. The work to be done is extensive and we should anticipate that you would have to spend at least £. . . . .

An old stone cauldron was situated in the left-hand side of the left-hand rear room and we would suggest that it is removed (unless you want to retain it as a feature).

The heating pipes in the cellar would seem to be lagged with asbestos. As you will be aware, there has been concern recently regarding this material. To remove it legally would cost a great deal of money but it can be 'encapsulated' (coated to prevent escape). This is the province of a specialist firm and we would not let your children play in the cellar until the treatment has been applied. (Cost unknown.)

The leak from beneath the front door step has previously been mentioned.

The room under the front living room had had the floor strengthened in places and we were led to believe that this was done to allow for heavy furniture that the owners had had some years ago.

As with the roof, we would advise that a woodworm and rot specialist inspects these areas and treats as necessary. The floor may also require further strengthening.

Our main conclusion on this area is that, if the new boiler were not in the basement drying off some of the moisture, then it would be extremely damp. This belief was supported by the damp timber by the walls, high moisture readings in the walls, spalling bricks, dampness on the floor together with efflorescence (powdery salts left behind after water has evaporated).

EXTERNAL AREAS

Gardens generally

The gardens were well kept and stocked but the structures (walls, greenhouse and sheds) displayed signs of years of neglect.

Rear garden

The defects noted were as follows:

(a) The greenhouse walls are badly bowed owing to soil pressure internally, the timber has been unpainted for years and woodrot has set in. If no repair work is carried out, it will collapse. We would anticipate having to spend approximately £. . . . . on repairs.
(b) The rear brick shed at the bottom of the garden needs clearing out and the roof repairing. A very approximate cost would be £. . . . .

(c) All garden walls are badly spalled and also need repointing. The approximate cost of patching up these walls would be £. . . . .

(d) The storehouses near the main building were badly maintained. Ivy had forced its way through the roof boarding in several places and the roof has badly rotted (wet rot probably). The flashings to the rear chimney were loose and this was part of the cause of wet rot in the boarding. The rear window frame was badly rotted. The buildings require a complete repointing. The doors and frames were starting to rot in places. The cost of remedial work would probably be in the region of £. . . . .but this is very much an assessment. The WC did not appear to work but we tested the Belfast sink and all seemed in order.

(e) The garage and side gate are badly rotted and rainwater was seen to be dripping in where the garage abuts the boundary wall with the neighbouring property. Once again an approximation of likely cost of repair would probably be £. . . . .

The rear paths which are mainly tarmacadam were in reasonable condition although sections of making good will be required eventually. The stone path which leads down the garden is uneven and needs relaying. Once again an approximation of likely cost of repair would probably be £. . . . .

Front garden

The faults were as follows:

(a) The front wall requires rebuilding. There is a lean on the wall of about 3 in in (3 ft). We noted that the neighbouring property had recently rebuilt their wall. Assessment of costs £. . . . .

(b) The right-hand boundary wall was badly spalled and there is evidence that some facing up of bricks has taken place in the past. This wall is also leaning and will require rebuilding eventually. Assessment of costs £. . . . .

(c) The old oil tank which is obsolete has never been removed and the trellis surrounding same is totally rotten. Assessment of cost of removal £. . . . .

(d) The left-hand wall has been built on a slope. No craftsman would build a wall like this. However, it does not seem to be in danger of falling down even though it has settled in the middle.

The front paths – refer to comments on rear paths above.

SUNDRIES

The engineer's, Coal Board, electrician's and plumber's reports are appended to the rear of this report and should be read in conjunction with same.

RATING AND PLANNING

We have not made any enquiries regarding rating or planning and have presumed that you or your solicitor will make these enquiries.

SCHEDULE OF COSTS

A schedule of approximate assessed costs is listed below for your reference:

1)	Loft access hatches	£. . . . .
2)	Roof ventilators	£. . . . .
3)	Work on chimneys	£. . . . .
4)	Renewal of sections of rainwater pipe	£. . . . .
5)	Renewing front gutters, fascia, etc.	£. . . . .
6)	Renewing vent pipe	£. . . . .
7)	DPC and plaster	£. . . . .
8)	Repointing	£. . . . .
9)	Mastic to windows and minor repairs	£. . . . .
10)	External painting	£. . . . .
11)	Electrical work	£. . . . .
12)	Water tank	£. . . . .
13)	Wall plaster (front bedroom)	£. . . . .
14)	Plaster in downstairs WC	£. . . . .
15)	New kitchen floor	£. . . . .
16)	Work in basement generally	£. . . . .
17)	Greenhouse	£. . . . .
18)	Shed	£. . . . .
19)	Repointing rear garden walls	£. . . . .
20)	Work to outhouses	£. . . . .
21)	Garage	£. . . . .
22)	Front wall	£. . . . .
23)	Side wall	£. . . . .
24)	Remove tank	£. . . . .

_____Approximate total   £. . . . .

THESE COSTS DO NOT INCLUDE ANY VALUE FOR ANY WOODWORM OR DRY ROT WORK OR ASBESTOS STABILIZING.

CONCLUSION

The above costs, which are considerable, reflect works being carried out by a small building contractor prior to the wage increase in June, 19. . . . . The dry rot and woodworm control work must be carried out very quickly to prevent further damage to the property. If minor works were carried out by individual tradesmen, it should be possible to reduce the cost slightly. It would also be possible to spread some of the non-essential work over a period of time.

However, if one takes the current asking price of £ . . . . . together with the predicted repair costs of £ . . . . . making a total of £ . . . . . , a newer property worth the same value would not offer the same facilities.

On the other hand we would anticipate you having several years of fairly hard work to put the house in order.

A. R. WILLIAMS MCIOB

For and on behalf of
Building Services (Technical)

Appendix B: Ninetails Avenue report

For

The Shambles,
72 Ninetails Avenue,
Often Narrows,
Near Tipham,
Tipfordshire

For

Mr F. Blofeld,
Wheatley Terraces,
(Off Satanists Walks),
Shepherd's Cottages,
Near Tipham,
Tipfordshire

and

Mr S. Tree,
70 Ninetails Avenue,
Often Narrows,
Near Tipham,
Tipfordshire

Building Services (Technical)
Old Ferry Offices,
The Square,
Tipham,
Tipfordshire

September, 19—

NOTE
THIS REPORT HAS BEEN INCLUDED TO ILLUSTRATE THE
EFFECTS THAT NEGLECT CAN HAVE ON A PROPERTY AND
ALSO THE NEED FOR CAUTION.

Although the following report is based upon a genuine survey, the names of any persons, companies, properties and the locations have been deliberately changed. Building Services (Technical) is a registered business name (Reg. no. 2175590).

In order to save space the various sketches and documents referred to in the text of the report have not been reproduced.

References to sketches in the report have been dashed out in order that there is no confusion with those referred to in the rest of the book. The sketch layout of Ninetails Avenue has been included at the beginning of Chapter 5.

The report has not been revised to allow for recent changes in the Building Regulations. Note in particular that recommended thicknesses of insulation have been changed since the original report was written.

BUILDING REPORT ON 72 NINETAILS AVENUE, OFTEN NARROWS, NEAR TIPHAM, TIPFORDSHIRE

INSTRUCTIONS, LIMITATIONS AND GENERAL PREAMBLE

We report that we carried out a survey at the above property on —— September, 19—— as instructed, in accordance with our letter of confirmation dated —— and the attached conditions of engagement (see copy at the rear of this report).

As instructed, the keys were obtained from Mr S. Tree of 70 Ninetails Avenue and returned immediately upon completion of our inspection.

This report shall be for the private and confidential use of our clients, Mr S. Tree and Mr F. Blofeld, for whom the report is undertaken and must not be reproduced in whole or in part or relied upon by a third party.

We will not be responsible to third parties who obtain, by any means, a copy of this report and act upon information contained therein.

In must also be understood that because the house was purchased prior to survey, this report should be considered as a preliminary schedule of defects (see also our conclusion). We at no time advised our clients to buy this property.

The report does not cover the condition of services (e.g. electrical, internal plumbing and heating). We understand that these systems are to be renewed in their entirety.

Whilst every care has been taken in completing our survey and report, the investigations have been non-destructive in nature and therefore we are unable to report upon matters which were concealed at the time of the inspection and the following assumptions have had to be made:

(a) That high alumina cement (HAC), concrete or calcium chloride additive or other deleterious materials or techniques were not used when the original property or any subsequent extensions were built.
(b) That any wall ties that exist are not perished.
(c) That your solicitor's or legal adviser's search will prove that the property is not subject any unusual covenants, Local Authority restrictions, or onerous restrictions imposed by others, encumbranes or outgoings and that good title can be shown.
(d) That future inspections of any hidden parts of the stucture which have not been inspected during this survey will not reveal material defects, or cause the surveyor to alter materially the indicative costings shown at the end of this report (if applicable).

No pressure, smoke or other tests have been applied to the drains.

A detailed inspection of some areas was not possible as furniture and the like were still in place at the time of inspection. Cupboards were found to be full of old pots, pans, clothing and the like.

Likewise it was not possible to fully examine the floors because there were fitted carpets or permanent coverings in all rooms. Corners of carpets were lifted where possible and isolated areas examined.

We have not inspected woodwork or other parts of the structure which are not uncovered, unexposed or are inaccessible. Neither have we removed insulation from between or laid over ceiling joists to inspect the hidden surfaces of the timber because such an undertaking constitutes extensive builder's work which does not form part of our conditions of engagement, nor have we inspected the wall cavities or wall ties. We are therefore unable to report that the property is free from defect. However, we have done our best to draw conclusions about the state of the property and construction from surface evidence visible at the time of our inspection. (*NB* Later in the report you will note that we found outbreaks of woodworm and woodrot in the basement and roof space.)

Whilst we have tried to amplify our descriptions when technical terms have been used in this report, we would also direct you to our standard cross-section through a typical building which details the major elements of a typical building and sketches bound into the rear of the report.

References to left and right are made facing the property standing in Ninetails Avenue.

THE PROPERTY GENERALLY

The property is a detached house, approximately a quarter-mile from a motorway junction.

Based upon local records, and information provided by Mr Tree, the property would appear to have been constructed about 60 years ago.

It comprises a cellar, hall, two living rooms, kitchen, store-room and rear porch to the ground floor and a landing, three bedrooms and a bathroom/WC upstairs.

The building is a traditional brick built structure with a slate roof. The main external walls are good quality red stock brickwork to the front and side. The rear walls are in common brickwork. Judging by the stretcher bond and observations made where a brick was missing, the walls are of cavity construction.

For additional reference, we have provided sketches and photocopies of the survey photographs at the rear of the report for your perusal (see appendix).

As you are familiar with the property, we have not provided superfluous information such as the distance from amenities or room sizes.

THE SURROUNDING AREA

From information provided by Mr Tree, we understand his house (70 Ninetails Avenue) was originally the caretaker's lodge for the former Goldfinger School of Painting for Girls which was demolished about ——— years ago after being deemed surplus to local requirements.

Mr Tree led us to believe that the future of the former school site is still uncertain. We have made enquiries at the local council offices and have been told that there are several proposals for the site. We believe that it would be worth requesting your local councillors and solicitor to pursue this matter further.

THE RECENT HISTORY OF PROPERTY

The following background information concerning the property was provided by Mr Tree and we have noted same for future record:

(a) According to Mr Tree the house has suffered from a planning blight for many years because of the possible effects of the proposed Often Narrows motorway link road.

(b) We were informed that Mr Bond's late father sold the property to the Local Authority some ————— years ago but the family remained as tenants until the property fell into extreme disrepair and became unfit for human habitation.

(c) We were led to believe that the house has been totally unoccupied for some four to five years and that, during that time, no regular inspections were carried out. During this period of non-occupation we understand that the central heating was switched off but that no attempt was made to drain down the water services. As a consequence, prior to and since repurchase these systems have suffered from frost damage on several occasions. In each case, following each subsequent thaw, major leaks on the plumbing systems serving the property were left unrepaired for weeks/months at a time.

(d) We were also informed that the property was offered and repurchased from the Local Authority some 18 months prior to our survey. We gather that the Local Authority valuer put it on the market at £. . . . . (well below market value in good condition).

(e) According to Mr Tree, no independent survey was carried out upon the property prior to purchase.

(f) We have also been led to believe that the house is at the moment owned jointly by Mr J. Bond, Mr S. Tree and Mr F. Blofeld.

(g) Following our inspection, we made enquiries at the Local Council Offices and would confirm that the link road route has now been finalized and that Ninetails Avenue will not be affected.

WET ROT, DRY ROT AND WOODWORM

Outbreaks of wet rot, dry rot (true *Serpula lacrymans*) and woodworm were noted in the cellar (see sketch plan). In the roof, the valley rafter seemed affected by rot. We were unable, because of plaster coverings, to determine the type.

Of the defects listed above, *Serpula lacrymans*, true dry rot, is the worst type. This fungus thrives in areas that are damp, badly ventilated and not too cold.

The cellar has all these conditions and hence a colony has established itself there. But dry rot, unlike other fungus, will search out timber in other parts of the house and has the power to force itself behind plaster and through brickwork. Once established dry rot is like a cancer. It will eat its way through an affected building with great rapidity, if allowed to do so.

It is therefore essential that reputable specialist eradication firms, who will give guarantees, are used to destroy the dry rot. This type of rot will not be deterred by anything other than the correct methods of destruction.

Any strands left untreated will recolonize the house and start to degrade the timbers once more.

We are of the opinion that these visible displays of rot and woodworm, *so far discovered* are only the tip of the iceberg. We believe that beneath the lath and plaster to the ground floor, in particular, there lurks further trouble.

This is because the dry rot will, by now, be working its way along hidden timbers and searching for other food sources.

The recommended treatment for dry rot is that all brickwork in dry rot areas be irrigated (holes drilled at close centres and a barrier formed by pumping in dry rot fluid) and the timbers near to the rot be cut out to at least 1 m from the last outbreak. The diseased timbers must then be destroyed.

Although we have indicated our findings on sketches at the rear of the report, we would recommend that the dry rot specialists recheck the whole house in detail, hacking inspection patches from plaster walls, lifting floorboards and removing slates, where necessary. We would also advise that they be called in quickly as true dry rot grows extremely quickly when conditions are favourable.

RISING DAMP

The damp proof course (DPC) in the walls did not seem totally effective. We would recommend that several reputable firms are asked for quotations for a new injection DPC to all external walls. If considered necessary, in the light of their reports, injection courses should also be installed in the internal walls. All firms should be asked to confirm that they offer 30-year guarantees on their workmanship and that the guarantees are backed by the fluid manufacturer. (As with the . . . . . system.)

Renewal of wall plaster adjacent to the new DPC is essential. The normal recommendation is that plaster be removed 1 m high above floor level and after the DPC has been injected an approved patent renovating plaster be used to make good. Standard plasters must not be used otherwise remaining salts in the wall may damage new decorations.

The installation of a new DPC will obviously mean that all the ground floor rooms will need a full decoration. However, as the house requires painting and decorating works carrying out, this will not create a problem, in our opinion.

THE MAIN STRUCTURE

The cellar generally

The floor to the cellar seemed in places to be breaking up and may require localized replacement. The cellar walls were covered with efflorescence. (This is a salt deposit brought in by damp and although not serious in itself indicates that a large amount of water has been evaporating in the area.)

The cellar area has been partially reviewed under the sections 'Wet rot, dry rot and woodworm', and 'Rising damp'; however, we would amplify as follows.

The rooms in the cellar were found to be damp, badly ventilated and strewn with rubbish and old furniture. (It is probable that the old furniture introduced the woodworm.) We understand from Mr Bond that his late father used to like visiting auctions. The cellar certainly reflects this. We found several old timber work benches (one affected by dry rot), old wooden packing cases, a set of old wooden stairs and many other miscellaneous timber items. We have stressed wood and timber quite deliberately.

Dry rot in particular likes areas like this in which to breed, and the more timber available, the faster it will spread. It would therefore be sensible to remove and destroy timbers not required, once the timber specialists have surveyed the house. We tested the walls, floor and ceiling in the cellar and found the moisture readings to be extremely high. Obviously, a large amount of the moisture will be rising from the surrounding ground but leaks from above have undoubtedly helped to dampen the basement area, particularly the lath and plaster of the ceiling. (The dampness will have assisted dry rot spread.)

The rooms above the cellar are supported on rolled steel joists (RSJs). Two of the main ones are marked on our sketch plans but there must also be others that are hidden from view by the lath and plaster ceiling. All the exposed beams in the cellar were found to be un-fireclad and very rusty. (Under modern regulations all steel should have half-hour fire resistance. This is achieved by cladding it with plasterboard or other approved material such as Vicuclad or by applying an intumescent paint.) (*NB* Intumescent paints foam up when heat is applied and thereby form a fire protective barrier.) When steel rusts, it expands to many times its original size. By so doing, it disrupts surrounding brickwork. The rust also weakens the beams.

It is essential that exposed and unexposed steel beams be replaced if necessary. If they are not replaced, the rust must be blasted away, and a rust inhibitor applied to all surfaces. The treated beams should then be adequately painted and fire protected. If there is any doubt concerning the strength of individual beams, then we would recommend replacement with new beams that will match the original beams in size, cross-section and weight per metre.

It is possible that some work will require Building Control approval prior to commencement and we would advise that the Local Authority should be consulted. You might also require the assistance of a consulting structural engineer to prove the adequacy of the present structure and/or repairs.

During our inspection, we discovered what appeared to be two old cellar drops. These have been sealed off but by so doing essential ventilation has been removed from the cellar.

Airbricks should be provided to the cellar in sufficient numbers to create good cross-ventilation (1500 mm^2 per metre run of external wall or 500 mm^2 per square metre of floor area which ever is greater is recommended).

We have incorporated a rough sketch of the cellar (see appendix). For ease of reference, we have divided the cellar area into three main rooms R1, R2 and R3.

R1

In R1 wet rot (assessed as Poria) was found on stored packing cases. Even though wet rot is not as serious as dry rot and will die out once the cause of the outbreak is removed (i.e. damp conditions), we would advise that all rubbish in this cellar be destroyed once the specialists have had a chance to re-examine the outbreak.

In this room, we found an old boiler but it seemed to have been disconnected. We would recommend that it is inspected by a competent plumber before re-installation.

R2

As indicated previously, a large outbreak of dry rot was noted. *Do not destroy the dry rot until the specialist has inspected*. Doing so will do little to hinder its growth and it might mislead the specialists as to the extent of the spread.

There has been a spillage on the floor. We were led to believe that it was antifreeze. This requires cleaning up. The cellar door is rotten and requires replacement.

R3

There was an old-style refrigerator in this room. Within the recent past, there have been several instances where children have been locked in and died inside this type of unit. We would recommend that it be disposed of, off site, after the door has been permanently removed.

Woodworm was found in a board which was covering an old cellar drop. As there is fresh frass (woodworm droppings) it is active, in our opinion.

As the beetles responsible fly around, we would anticipate other outbreaks becoming apparent once old carpets, boxes, etc., are removed and the area can be re-inspected.

Cellar staircase

The concrete and brick staircase leading from the cellar to the rear garden was found to be in poor condition. The concrete steps need attention and the brick retaining walls need rebuilding in places. Moss should be cleaned off all walls and steps, using an approved moss killer.

There is a steel beam above the cellar door which supports the house above. The wall above the steel beam is badly cracked because of rust expansion. This beam needs cutting out and replacing with a concrete lintel.

The 'outrigger' wall above has bulged because of the steel expansion and requires rebuilding.

The vertical crack in the cellar wall needs cutting out and the brickwork 'stitching' by a competent bricklayer.

The main pitched roof

Where practicable, the interior of the main pitched roof has been examined but the inspection of the roof covering has been made from ground level only.

The main roof is of traditional rafter and purlin construction and is covered in slates. The roof did not have a sarking felt layer beneath the slates, merely a rendering, which is not uncommon with a property of this age. (Sarking felt is a secondary layer normally incorporated into new properties to prevent water penetration if a slate is dislodged.)

The main roof seemed reasonably sound when inspected visually from the outside but there were several slipped or broken slates on the front, side and rear elevations. Tingles were also noted. (Tingles are pieces of metal inserted under a slate to prevent it slipping.) We suspect that the roof coverings will need stripping as the slate nails seem to be perishing. As an absolute minimum, the areas with missing slates need matching slates inserting. This work should be carried out by a competent roofer. The ridge and hip tiles need removing, rebedding and repointing.

We would advise that if the property is to have major remedial work carried out that the roof slates be stripped off, sarking felt placed in position and then the slates replaced. If the roof covering is totally renewed, it would be unwise to replace the slates with a cheaper substitute such as Messrs type X or Messrs Type Y tiles. These are much thicker and therefore heavier than the slates and could overload the roof timbers.

We would also bring to your attention that the main roof has a valley on the front elevation. (See our notes on the photographs contained in appendix.)

Whilst being totally acceptable forms of construction, valleys can create weaknesses in a roof. Once the leadwork perishes (which does take a considerable number of years), dampness can enter the roof space. Bearing in mind the age of the property, and the dampness detected internally, we would advise that this valley be inspected by a roofer and/or plumber. It may also be necessary to replace the valley rafter with treated timber as there are signs of rot apparent. (If dry rot (*Serpula*) is diagnosed by the rot specialist, then his advice must be requested before proceeding.)

We gained access to the loft through a trapdoor in the bathroom. The trapdoor seemed of reasonable size.

The loft was dirty and contained a large assortment of boxes and other household junk.

Naturally, most of the dirt was on the horizontal surfaces and it was therefore not possible to fully inspect the ceiling joists. The loft floor generally was insulated with approximately 1 in (25 mm) of glassfibre insulation but some areas had newer quilt layers laid over the joists (not between them). As this method of laying quilt can create access problems, we would recommend that these areas be relaid. The areas with only 25 mm quilt should be upgraded. The normal accepted standard for roof insulation is now 100 mm but we anticipate that insulation standards may well be increased in the near future.

However, we believe that it would be unwise to lay insulation that is thicker than the depth of the joists in the loft because doing so could easily obscure the roof timbers and make access hazardous. (The joists in the main seemed to be approximately 100 mm deep.)

When the new insulation is laid, it should be kept back from the edges of the roof so that the ventilation from the eaves is not obstructed.

There was a strong smell coming from within the loft void and large numbers of heavy cobwebs and flies were apparent. We were not able to fully inspect the roof space owing to the clutter. Also the swarms of flies attracted by the lamp dangerously obscured visibility. We suspected that there may be dead birds in the loft space which will require removal.

This whole area obviously needs further investigation once it has had the rubbish removed.

After comparing the roof timber sizes with modern tables, we came to the conclusion that by modern standards the timbers were undersized. However, the roof has no signs of distress and we would therefore conclude that the roof construction is sound.

The purlins are supported on dwarf walls (a purlin is a main ring beam which holds up the roof). We tested the timbers in several locations with a moisture meter and generally speaking found the readings were within acceptable limits. The main exception to this was around the chimney stacks. There was evidence of water penetration on the render surrounding the stacks within the loft space. The structural members around the chimneys and the gutter boards showed high moisture readings which would lead us to believe that the flashings and soakers required attention or replacement (soakers and flashings are purpose made pieces of lead inserted to fill in a junction between the chimney and the roof). A header tank was noted by the left trapdoor but it was empty and the piping was not connected. The inside of the tank showed definite signs of corrosion.

Note
Owing to difficulty of entry, the section on the left-hand side of the loft could not be inspected at close quarters and neither could the far side of the chimney stack.

Chimneys (2 no.)

There were two chimneys. When inspected from ground level with binoculars it appeared that the stacks needed fully repointing. It would also be advisable to re-flaunch the tops of the chimneys. (Flaunching is the cement and sand topping to the chimney.) The lead flashings also require attention. We would also advise that the chimneys be swept as falls of soot were noted in the fireplaces.

Main walls of house

No examination has been made of the foundations because this would require excavating inspection pits. The back door step has subsided and needs removing and resetting in position.

There was evidence of a damp proof course (DPC) in the external walls. This was of an old-fashioned construction (i.e. layers of tile or slate). The DPC was generally 150 mm (6 in) above ground level which is the recommended minimum. (As we understand it, 150 mm is supposed to be the maximum height that splashing rain can leap up off a paved surface in very wet weather.) However, moisture meter tests internally and externally indicate that the DPC is not totally effective. As previously stated, we would advise that this be inspected by a firm of damp proof specialists and a quotation be obtained for an injection DPC. All brickwork elevations need raking out and repointing. By the driveway there is an old steel beam built into the wall. As it is in danger of rusting, in a manner similar to the one down the cellar steps, and disrupting the brickwork, it should be cut out and replaced by a reinforced concrete lintel. The artificial stone cills to windows were flaking in places. These should be repaired using a coloured mortar repair technique.

When we checked the walls with a large spirit level, we found that the walls in the main were plumb although the cills indicated some slight settlement. This could be caused in part by the very heavy traffic from Ninetails Avenue vibrating the house on its foundations over a number of years. The rear 'outrigger,' as indicated previously, has been disrupted by a corroding steel joist in the cellar area and needs rebuilding in places. A brick is missing adjacent to the kitchen porch and must be replaced.

Note
Recent reports have indicated that cavity wall tie corrosion in the UK is becoming a major problem. With a property of this age, we would anticipate that some or all of the wall ties in cavities will require replacement. However, no tests have been made on the ties as this is beyond the scope of our conditions of engagement.

Drainage

No manhole was visible at the time of inspection, so we were unable to carry out any drainage inspections. From discussions with Mr Tree, we were led to believe that the nearest manhole that serves the system is adjacent to the demolished school buildings some 80–90 ft away from the property. It is worth pointing out that in the event of a blockage the system could be difficult to rod. Several gulleys need cleaning out as they are full of leaves, dirt and sapling trees.

After considering the location of two major trees and the fact that the rear paving. has subsided, we are of the opinion that the drains have been disrupted by tree roots. The running water has then washed away the earth below the paved areas. This running water will also not help the conditions in the cellar. (See later regarding trees.)

Windows, doors and timberwork

The external woodwork is in need of repainting. As we only carry a 3 m ladder it was not possible to inspect the high level woodwork but based upon the conditions found below we would anticipate that it will need some renewal. All external woodwork needs the paint burning off down to sound wood, defective sections of timber should be cut out and replaced with treated timbers of a similar size and then all the woodwork should be 'knotted', primed, undercoated and glossed with a good quality painting system. All the window stays inspected seemed antiquated and in need of renewal.

The door frames will need the bottom 6 in (150 mm) of the frames removing and new timber splicing in or the frames renewing completely. The front and back doors need replacing. It will probably be possible to save the windows but some splicing in of new timber will undoubtedly be necessary in places.

Gutters, downpipes and leadwork

In the main the high level gutters have been replaced with PVC but the downpipes are still in cast iron. Moss growth was apparent on the walls behind several pipes which proves that they are leaking and have been discharging water onto the house walls for some considerable time. Upon further inspection we found that all the cast iron pipework was badly rusted owing to the fact that they have not been painted where hidden from view. In our opinion, all the cast iron pipes require replacement with new PVC pipes.

The gutters over the rear left-hand bedroom (viewed from front) need refixing as they are hanging loose.

On the front pediment there is a short length of gutter on the left-hand side. This is drained by a short length of bent pipe. We would advise that the gutter and pipe be checked carefully for blockages.

As previously mentioned elsewhere, because of the age of the property, and bearing in mind the possible perished leadwork to the valley, we would advise that flashings and soakers be checked and replaced, if necessary, by a competent plumber. The half brick left lying in the valley gutter should also be removed.

INTERIOR GENERALLY

Gas supply

No test was carried out. Services beyond the scope of this report.

Heating installation

No test was applied. Services beyond the scope of this report but note previous comments concerning extensive leaking.

Plumbing

No test was applied. Services beyond the scope of this report but note previous comments concerning extensive leaking.

The bath was badly stained and the other fittings looked old. A large number of old lead pipes were noted within the dwelling. As lead pipework can be a health hazard, it would be advisable that this pipework be replaced with copper/stainless steel or patent plastic piping.

Electrical installation

No test was applied. The electrical system generally was found to be suspect and inadequate. We would recommend a complete rewire. The meters are in the basement, and as these seem rather antiquated, we would advise that they be inspected by your local electricity board and renewed as necessary.

Floors

All floors were covered by fitted carpet or stuck finishes. It was therefore only possible to inspect in isolated areas. In the main all floors seemed to be of timber construction. Owing to the extensive leaks, and the dry rot in the lower floor area, we would anticipate rot attacks being discovered in many places.

Plastering and decoration

The property internally was generally in a bad state of repair. We would anticipate that most of the wall plaster will require renewal once the innumerable layers of wallpaper have been stripped away. This will certainly be necessary where water damage has occurred. The ceilings seemed in slightly better condition but, once again, large areas have been subject to the effects of running water.

Internally, the house requires complete redecoration. Externally, all fascias, windows, doors, etc., require burning off and a complete retreatment of either paint or varnish to match existing.

CELLAR

See before.

GROUND FLOOR

Rear porch

The floor was rotten. The roof, windows and woodwork need replacing. In reality the whole structure should be pulled down and renewed.

Front porch

The tiles seemed to have settled. Ditto as for rear porch.

Hall

A small piece of glass is missing from the leaded light. There is also a cracked pane. Moss and mould noted on wallpaper adjacent to leaks.
   (See before regarding comments on rising damp, rot, water damage, plaster, decorations, etc. See also sketch ——.)

Lounge 1

The flat roof to the bay needs attention. Standing water could be seen on the roof and there was evidence of water percolation on the ceiling. The roof should be cleaned off and outlets cleared. If the dampness persists, the lead roofing should be replaced.
   (See before regarding comments on rising damp, rot, water damage, plaster, decorations, etc. See also sketch ——.)

Lounge 2

All as for lounge 1.

Kitchen

All the kitchen units were in extremely poor condition and, in our opinion, could not be made fit for human use. Rat droppings were noted in the units. Addled plaster was noted on the kitchen wall and water runs by the chimney breast.

(See before regarding comments on rising damp, rot, water damage, plaster, decorations, etc. See also sketch ———.)

FIRST FLOOR

Bathroom

There was a large crack in the wall plaster caused by the damage to the external wall. (See details concerning rusted beam in the cellar.)

The bathroom plaster had been textured into various shapes. Unless one wishes to preserve this feature, then the wall plaster will have to be knocked off and replastered. The bath was badly stained and all the fittings were antiquated. They will need stripping out and renewing. It was not possible to test the fittings as the water had been turned off. Wood rot was found in the floor and skirting. Rat droppings were found in the cupboard. The tank in the cupboard is badly rusted and piping leading to the header tank above is missing. (See before regarding comments on rising damp, rot, water damage, plaster, decorations, etc. See also sketch ———.)

Landing

Plaster is missing below the valley rafter. Rot found in the valley rafter. (See before regarding comments on rising damp, rot, water damage, plaster, decorations, etc. See also sketch ———.)

Bedroom 1

Corrosion and leakage noted on radiator. (See before regarding comments on rising damp, rot, water damage, plaster, decorations, etc. See also sketch ———.)

Bedroom 2

This room has an oriel window. The pebbledash to the oriel externally needs renewing. The wall plaster noted badly cracked in cupboard. This could once again be caused by wall movement owing to condition of the rusting beam in the cellar. (See before regarding comments on rising damp, rot, water damage, plaster, decorations, etc. See also sketch ———.)

Bedroom 3

(See before regarding comments on rising damp, rot, water damage, plaster, decoration, etc. See also sketch ———.)

EXTERNALLY

Rear gardens

The garden is overgrown but is fairly large by modern standards. All the *in situ* concrete paths require complete renewal.

As previously mentioned, the subsoil has been washed away by leaking drains and now all the paths are irregular and cracked.

There are two large trees in the rear garden close to the right-hand corner of the house. Their presence concerns us because their roots could well damage the foundations of the house.

We believe that they should be felled *but only after checking with the local authority that no tree preservation orders exist.*

The elderberry growing alongside the house should also be cut down and the roots grubbed up. If this is not done, the trunk will exert pressure on the cellar walls and eventually cause collapse.

Front and side garden

Like the back garden, the front is overgrown. A tree has started to push the wall over at the front. The tree should be cut down (after checks with the Local Authority) and the wall rebuilt. All paths need renewing as per the rear garden.

RATING, PLANNING, BUILDING CONTROL AND COAL BOARD

We have not made any enquiries regarding rating, planning, building Control or Coal Board surveys.

From your instructions, we were led to believe that your solicitor was making these enquiries.

CONCLUSIONS

This house requires a great deal of attention. If enough money is spent on it, it could have potential. It was obviously a fine building when it was originally constructed and only time and lack of care have allowed it to deteriorate.

We would stress that it would be unwise to spend money on the property without carrying out extensive costings (which are beyond our present brief) and consulting a local estate agent/experienced valuer about likely resale values. We believe that establishing a value for this house once the defects have been rectified could be difficult until the future of usage of the adjacent land (former school site) is known beyond doubt.

However, if Ninetails Avenue were not to carry the heavy traffic that it does at present, then we would anticipate that the value of all the properties in the area would rise substantially in any case.

Under normal circumstances, we would attempt to give some idea of cost of the remedial works required to put this house in order. In this case, more investigation and assessment of rot damage will be needed before a reasonable costing could be prepared.

We have recommended that dry rot specialists/DPC specialists be asked to make preliminary investigations and to give likely costings for their work and to provide details of likely builders' work. Once this is known, then costings could be prepared with some degree of confidence. Should you require our assistance in this matter, please feel free to contact us.

A. R. WILLIAMS MCIOB

For and on behalf of
Building Services (Technical)

Appendix C: Acacia Avenue report

For

3 Acacia Avenue,
Shepherd's Cottages,
Near Little Bodworth,
Tipfordshire

For

Mr and Mrs Mayo,
Dunromin,
Pentangle Road,
Often Narrows,
Near Tipham,
Tipfordshire

Building Services (Technical)
Old Ferry Offices,
The Square,
Tipham,
Tipfordshire

June, 19—

NOTE
THIS REPORT HAS BEEN INCLUDED TO ILLUSTRATE THAT THE SURVEYOR HAS TO BE ON HIS/HER GUARD. ON FIRST SIGHT THIS PROPERTY LOOKED WELL MAINTAINED.

Although the following report is based upon a genuine survey, the names of any persons, companies, properties and the locations have been deliberately changed. Building Services (Technical) is a registered business name (Reg. no. 2175590).

The various sketches and documents referred to in the text of the main report have not been reproduced.

References to sketches in the report have been dashed out in order that there is no confusion with those referred to in the rest of the book.

The report has not been revised to allow for recent changes in the Building Regulations. Note in particular that recommended thicknesses of insulation have been changed since the original report was written.

BUILDING REPORT ON 3 ACACIA AVENUE, SHEPHERD'S COTTAGES, NEAR LITTLE BODWORTH, TIPFORDSHIRE

INSTRUCTIONS, LIMITATIONS AND GENERAL PREAMBLE

We report that we carried out a survey at the above property on. . . . June, 19. . . . as per our letter of confirmation dated. . . . June, 19. . . . and our standard conditions of engagement (as previously forwarded, see copy in appendix).

This report shall be for the private and confidential use of our client for whom the report is undertaken and must not be reproduced in whole or in part or relied upon by a third party.

We will not be responsible to third parties who obtain, by any means, a copy of this report and act upon information contained therein.

Whilst every care has been taken in completing our survey and report, the investigations have been non-destructive in nature and therefore we are unable to report upon matters which are concealed at the time of the inspection and the following assumptions have had to be made:

(a) That high alumina cement (HAC), concrete or calcium chloride additive or other deleterious materials or techniques were not used when the original property or any subsequent extensions were built.
(b) That any wall ties that exist are not perished.
(c) That your solicitor's or legal adviser's search will prove that the property is not subject to any unusual covenants, Local Authority restrictions, or onerous restrictions imposed by others, encumbrances or outgoings and that good title can be shown.
(d) That future inspections of any hidden parts of the structure which have not been inspected during this survey will not reveal material defects, or cause the surveyor to alter materially the indicative costings shown at the end of this report (if applicable).

We have not inspected woodwork or other parts of the structure which are not uncovered, unexposed or are inaccessible and we are therefore unable to report that the property is free from defect; however, we have done our best to draw conclusions about the construction from surface evidence visible at the time of our inspection. We have not inspected the wall cavities or wall ties.

It was not possible to fully examine the floors because there were fitted carpets or permanent coverings in all rooms. Corners of carpet were lifted where possible and isolated areas examined.

No pressure, smoke or other tests have been applied to the drains.

As promised, we have included indicative costings to cover likely remedial works. However, we would stress that these are only intended to be an approximate guide and not an exact cost. They are given so that you understand the likely implications of the purchase in approximate monetary terms. As the costings are based upon average building prices, assumed specifications and assumed areas, some of the estimates may be high and some may be low. It is essential therefore that competitive estimates are obtained from several local builders prior to instructing work to be carried out. Our costings should only be treated as a guide, current at the time of preparing this report with no allowance made for value added tax.

The survey commenced at approximately 1.30 p.m. The weather conditions were fine but windy.

References to left and right are made facing the property standing in Acacia Avenue (i.e. the left adjoins number 1).

SPECIFIC INSTRUCTIONS

We were particularly asked to inspect the damp proof course and test its effectiveness. We understand that the current owners claim that an injection damp course was installed by the previous owner.

THE PROPERTY GENERALLY

The property is a semi-detached house, directly opposite the junction with Cyclops Road. From discussions with the present owner, we understand that the original property was built just after the last war. It has subsequently been extended at the side and rear.

The building is a traditional brick built structure with a tile roof. One side of the roof was hipped and the hips were covered with purpose-made tiles. The original structure has render and spar dashed walls with facing brick feature bands. The rear and side walls of the original structure are similar except that they are only covered in smooth rendering. The areas below the cill level of the ground floor windows are also facing brick.

The property comprised an open front porch, hall, front living room, rear living room, kitchen and dining area to the ground floor and, landing, three bedrooms and shower room/WC upstairs.

The walls of the original house appear to be of solid construction. The two new extensions are of cavity construction.

Behind the feature boarding to the front pediment of the original house (these walls were inspected in the loft space), the construction is of half brick and one brick construction (i.e. not cavity). It is also worth noting that this property has a corner window which although attractive can be an inherent source of weakness but no signs of stress are apparent at the time of inspection.

As you are reasonably familiar with the property and area, we have not provided superfluous information such as the location of schools in the area etc. We would, however, remind you that the adjoining property, number 1, seems to have been unoccupied for a long time. For reference we have provided photocopies of the survey photographs at the rear of the report for your perusal. (*Note* Photographs not reproduced.)

We would remind you that no test has been made on the wall ties (if applicable) as this would require drilling the walls and inspecting the cavities with an endoscope which is beyond the terms of our engagement. Should you require such tests carrying out we will instruct a specialist company on your behalf.

THE MAIN STRUCTURE

The main pitched roof

Long ladders and roof ladders have not been used for inspection.

Where practicable, the interior of the main roofs have been examined but the inspection of the roof covering has been made from ground level only, using binoculars.

On the front elevation there was one area which could not be visually inspected from ground level because of the roof intersection, but we have no reason to doubt that all the tiling was of a similar condition. (*NB* The limitations to the extent of the survey recorded.)

The main roof was of traditional rafter and purlin construction and was covered in tiles. The roof did not have a sarking felt layer beneath the tiles, merely a cement and sand render layer, which is not uncommon with a property of this age. (Sarking felt is a secondary layer normally incorporated into new properties to prevent water penetration if a tile is dislodged.)

The main tiled roof seemed reasonably sound when inspected visually from the outside but there were several slipped or broken tiles on the front, side and rear elevations. These areas need matching tiles inserting by a competent roofer. The ridge tiles adjacent to the front pediment also need rebedding. The purpose-made hip tiles need repointing in places. There were areas of the roof near the rear chimney stack which seemed 'wavy'. We came to the conclusion that this area had probably been like that since the day it was built but it would be worth having the roofer inspect it during his visit and due account be taken of his recommendations.

We would also bring to your attention that the main roof has a valley and secret gutter on the front elevation (see notes on the photocopies of the photographs at the rear of the report for location, Photograph —— in particular).

Whilst being totally acceptable forms of construction, valley and secret gutters can create weaknesses in a roof. Once the leadwork perishes (which does take a considerable number of years), dampness can enter the roof space. We would advise that this be checked out by your roofer or plumber because the house is now over 50 years old and we would envisage that the lead will be well past its best.

Delamination of the plywood spandrel infilling to the roof over the front door was noted. (See Photograph ——.)

We gained access to the loft through a trapdoor on the upstairs landing. Whilst we did so with ease, we could envisage someone with a more stocky frame having problems as the trap was somewhat undersized.

The floor of the loft was dirty. This comment is not intended to reflect on the existing owners. What we found on the loft floor was the accumulated dirt left by years of atmospheric pollution. Naturally, most of the dirt was on the horizontal surfaces and it was therefore not possible to fully inspect the ceiling joists. The loft floor was unboarded and insulated with approximately 1 in (25 mm) of glassfibre.

The normal accepted standard for loft insulation is now 4 in (100 mm) but there are recommendations that 6 in (150 mm) of glassfibre be adopted as a new standard.

Subject to depth of joists in the loft and increasing roof ventilation, we would recommend increasing the insulation standards. The glassfibre should not be thicker than the joists otherwise it might obscure the timbers from view and make movement in the loft dangerous. At the time that the additional insulation is incorporated, patent roof ventilators and eaves ventilators to comply with modern regulations should be provided. Our reason for suggesting improved ventilation is that condensation is becoming a major problem in domestic properties. Improved insulation and lack of draughts in modern or improved properties allows the air to hold higher levels of moisture than was previously possible. At night, or in cold weather, this surplus moisture can condense out of the moisture laden air, especially in unventilated loft spaces and on cold walls. For the roof slopes on this property the unrestricted airflow gaps should be as indicated (see sketch at end of report).

The main timber sizes were as follows:

Type	Size (mm)	Spacing (mm)	Dist./supports (m)
Rafters	70 × 60	350	1.80
Ceiling joists	50 × 80		Varied
Purlins	70 × 170		Varied

After comparing the sizes with modern tables, we came to the conclusion that by modern standards the timbers were undersized. However, the roof has no signs of distress (other than the wavy patch previously mentioned) and we would therefore conclude that the roof construction is sound.

It is worth noting, however, that the purlins were supported on timber posts (and not the usual dwarf walls). (A purlin is a main ring beam which holds up the roof.) The feet of the two posts did not appear to have any obvious lateral support and, whilst one has to accept that the roof must be reasonably well constructed (otherwise it would have collapsed long ago), we would suggest that some cross-bracing to these posts might be desirable. We would also suggest that the posts be inspected by a competent joiner and strengthened if necessary. We tested the timbers in several locations with a moisture meter and generally speaking found the readings were within acceptable limits. The main exception was around the chimney stacks. There was evidence of water penetration on the rear stack. The structural members around the chimneys and the gutter boards showed high moisture readings which would lead us to believe that the flashings and soakers required replacement. (These are purpose-made pieces of lead inserted to fill in a junction between the chimney and the roof.) In places, the party wall between 1 and 3 had bricks missing. These should be blocked as the openings are a fire hazard.

The loft space did not appear to have any service tanks.

Two pitched roofs to the new extensions

The new kitchen/lounge extensions were covered with Marley Modern or similar interlocking tiles.

It was not possible to enter the roof voids to these roofs because there were no trapdoors. We therefore cannot comment on the condition of these loft areas. If you do purchase this property, then the condition of these areas must be at your risk.

The pitches and ventilation of both lower roofs seemed adequate. Two modern roof ventilators were noted on the rear extension and airbricks were noted on the adjoining walls. One or two tiles on the side elevation need relaying as they are standing proud of the main roof.

Where the neighbour's garage, at number 5, abuts the side extension, there is a hole in the felt roof covering. This could allow water to affect the new extension if not rectified by the neighbour.

Chimneys (2 no.)

There were two chimneys. When inspected from ground level with binoculars, it appeared that the joints were badly smeared and that the top four or five courses had not been pointed at all. We believe that the stacks need fully repointing. At the side of the front chimney stack there was a badly chipped brick visible which needed cutting out and replacing. It would also be advisable to re-flaunch the tops of the chimneys. (Flaunching is the cement and sand topping to the chimney.) The lead flashings and soakers also require attention as previously mentioned.

Main walls of house

No examination has been made of the foundations because this would require excavating inspection pits.

Damp proof course

There was evidence of the original damp proof course (DPC) in the external walls.

The brickwork below DPC level to the side elevation was covered in moss. This would indicate that the walls below the DPC are very damp.

We could find no evidence of an injection damp course being installed.

(Normally, if an injection course has been installed there is usually evidence in the form of a regular band of small holes or mortar patches 150 mm (6 in) above ground level. If an injection course is present, then the work must have been carried out internally and not from the outside, which is the norm in our experience.)

The original DPC was generally 150 mm (6 in) above ground level which is the recommended minimum. However, moisture meter tests internally and externally indicate that the DPC is not totally effective.

We would advise that this be inspected by a firm of damp proof specialists and a quotation be obtained for renewing the injection DPC prior to purchase.

External render and spar dash areas

Inside the front porch the wall surfaces are smooth rendered. A section of this render has fallen away from the brick backing. Deep cracks were noted on corners and the render surfaces were crazed. When the render was tapped it seemed hollow and would indicate that most of the render in the porch area is loose and should be renewed.

The main rendered areas of the side walls, within easy reach of the ladder, were also tapped. They also rang hollow and would indicate that most of this renderwork may well need renewal.

The reason for this is that when render becomes old it tends to crack and loosen its bond on the brick backing. The pockets formed then fill with water which eventually dampen the brickwork and can cause problems with plasterwork and timber internally.

Normal good practice would dictate that the old render be hacked back to sound surfaces, then an approved mortar mix be applied in layers.

It was not possible to check the front rendering in the same way because of the spar dash. However, judging by the bulging overlap under the rainwater pipe at the junction of number 1 and number 3, we believe that in this case the new spar dash has been applied over the top of the existing render without any preparatory remedial work being carried out.

The spar work, in our opinion, is purely a cosmetic exercise and probably covers many defects. It would seem that the existing render has been coated with a PVA adhesive and the spar dash thrown on whilst the adhesive layer was still wet. We have no doubt that within a very short space of time the bulk of the spar dash will fall away (as it is already doing).

Where the rendered areas meet the brickwork there is no bellmouth to throw rain water clear of the wall. Without a bellmouth water may run down the rendered surfaces, lodge in the junction and eventually dampen the plaster and timberwork internally. We would recommend that if and when the rendered work is renewed that a bellmouth be incorporated (see sketch).

External walls generally

When we checked the walls with a long spirit level, we found that the walls in the main were plumb. (On the spar dashed sections it was difficult to make precise readings.)

On the front wall of the side extension a 100 mm diameter hole was visible. There was no protective cover over this aperture and the room inside was exposed to the elements. Upon enquiring with the present owners, we were informed that the hole was originally intended as an outlet for a tumble dryer. The hole should either be used for this purpose or blocked off.

Drainage

No manhole was visible at the time of inspection, so we were unable to carry out any drainage inspections. According to the present owner, the nearest manhole that serves the system is in an adjacent garden. Apparently the system has had to be cleared and rodded by Messrs . . . . . several times in the last few years. We would conclude that there could be problems with the drainage system created by incorrect or insufficient 'falls'.

Several gullies need cleaning out as they are full of leaves and dirt.

The gully by the new side extension adjacent to number 5 requires the waste pipes being taken down below the cover level as the present position of the piping does not comply with modern regulations. The pipe should discharge into the trap of the gully and not above the grating as it does at the moment. The reason for this requirement is that water can splash up the walls and create damp problems.

Windows, doors and timberwork

All the windows in the property have been replaced with stained softwood single glazed windows. Judging by the condition, we would estimate that they are no more than five years old. Although they are sound, it is also obvious that only one coat of stain preservative has been applied. Therefore, the windows need re-treatment urgently.

As previously mentioned, the first floor window to the box room is a corner window. Whilst this can look an attractive feature, it could be a source of future problems, because the corner of a building is one of its strong points and a corner window, badly replaced, could weaken a building. If the property is purchased, we would advise that the area be checked regularly for any signs of movement. This is especially important considering the recent change of windows.

The woodwork to the doors, door frames and fascias, etc., in the main seemed sound. Screw holes noted in the French windows of the side extension would indicate that the doors have been reversed at sometime in the past. No mastic was apparent around doors and window frames. It is essential that this be rectified.

Gutters and downpipes

In the main the high level gutters have been replaced with PVC but the downpipes are still cast iron. Moss growth was apparent on the front wall adjacent to the junction between number 1 and number 3 which proves that this pipe has been leaking and discharging water onto the house wall for some time. Upon further inspection we found that all the cast iron pipework was badly rusted owing to the fact that the backs of pipes had not been painted where hidden from view. Adjacent to the front door there is a short length of cast iron gutter. This does not have proper stops on each end. Instead, rounded pieces of wood have been inserted. Whilst being far from satisfactory in the first place, now that the wooden stops have weathered, it is possible to see daylight around the bases of same. No doubt in wet weather, the rain cascades onto the ground below. In our opinion, all cast iron pipes and gutters require replacement with new PVC.

The extension generally had PVC gutters and downpipes which seemed in reasonable condition.

INTERIOR GENERALLY

Gas supply

The present owner indicated that a gas service has been laid onto the property. There did not appear to be a meter. No test was carried out.

Heating installation

The whole house was warmed by a series of electric storage radiators. Being summer these were inoperative but we had no reason to doubt that they worked. No test was applied. In the rear lounge there is an imitation log fire which the present owner will be leaving.

Plumbing

The sanitary fittings were tested visually by turning on taps and flushing toilets. All seemed in order. It is worth noting that the house has no bath, merely a shower.

Externally in the back garden there is an outside tap which seems to have been boxed in. No lagging was apparent. The boxing requires painting and the pipes lagging. There is a short stub pipe projecting from the wall of the new extension with a blank off cap fitted. We could not work out the purpose and the present owners did not know. The pipe is unlagged. This should be rectified.

Electrical installation

In the roof space there was evidence of cables that had been badly terminated. Covers were missing to spotlights in the kitchen and should be replaced for safety reasons.

The electric meters were in the hall. A white meter appeared to serve the heating system mentioned above.

Wires noted projecting in the rear lounge. These are for wall lights, according to the present owner. We gather that final connection to the power supply has yet to be made.

The normal household system was controlled by a six-way fuse board. Although it was obvious that some cabling was PVC, we were unable to ascertain whether the system has been totally renewed. Prior to purchase, we would advise that an approved electrician be asked to check the system. Such checks may well prove that there are still some rubber insulated cables in existence which tend to perish after 25 years' life. Any rubber insulated cable should be replaced by PVC covered cable and comply with the IEE regulations. The other reason for caution is that it is obvious that the electrical circuits have been altered at regular intervals by previous owners.

The points were as follows.

	Light points	Sockets	Other
Porch	—	—	—
Hall	1		
Living room (front)	1	2 doubles	Heater, TV point

Living room (rear)	2	5 doubles	TV point, heater, telephone
Kitchen	2	3 doubles	—
Dining room	2	4 doubles	TV point, heater
Landing	1		
Bed. 1 (small)	1	1 double	—
Bed. 2 (front)	1	2 doubles	—
Bed. 3 (rear)	1	1 double	—
Shower room	1	—	Towel rail

Floors

All floors were covered by fitted carpet or stuck finishes. It was therefore only possible to inspect in isolated areas.

Judging by the fact that the ground floor screed is burying the bottom edge of the skirtings in places and the high moisture readings on the ground floor and skirtings we came to the conclusion that the person who had laid the screed had merely done a cosmetic cover up of existing defective floors.

Decoration

The property generally was in a reasonable state of decoration internally. Externally, all fascias, windows, doors, etc., require a complete re-treatment of either paint or varnish to match existing.

Most walls internally are papered or pine boarded. The ceilings are generally of the Artex type. We suspect that as the ceilings are lath and plaster construction that the Artex work and the pine boarding has been applied to disguise the cracked or defective surfaces below. *Note* In the kitchen, the ceiling in one part has been damaged. According to the current owner, this was caused by the cork off a champagne bottle hitting the ceiling and dislodging the finish.

Note If an injection DPC were installed, this would largely destroy the decorations on the ground floor because of the necessity to replaster the first metre of wall following the insertion. If the plaster is not renewed in patent renovating plaster or waterproof cement and sand, then the DPC company's guarantee will be void and damp will re-establish itself owing to the salts that will still be present in the old plaster.

GROUND FLOOR

Porch

The entrance porch has a quarry tile floor which seems in good condition. Note comments previously on cracked render.

Hall

Note previous comments on floor. (For details regarding decoration, electrics, plumbing and the like see previous comments.)

Living room (front)

Note comments on floor. (Ditto.)

Living room (rear)

Note comments on floor. (Ditto.)

Kitchen/dining room

The kitchen is plumbed for a dishwasher and washing machine. Under one unit there is a cracked tile. All kitchen units seem in reasonable order. The walls are partially papered, tiled and pine boarded.

FIRST FLOOR

Shower room

The ceiling and walls were pine boarded. Inside the shower, the walls were tiled. We tested the fittings by running water and all seemed in order. There is no bath, only a shower fitting, which is unusual. We did not test the electric shower.

The towel rail and heater in the bathroom are electric. The cabling disappeared behind the wall panelling. There was no obvious way of disconnecting same. We would advise that this cable be inspected by an approved electrician. The shower cubicle door was stiff. (For details regarding decoration, electrics, plumbing, floors and the like see previous comments.)

Landing

The landing window was very high for operation by a small person. No other comments. (For details regarding decoration, electrics, etc., see before.)

Front box room (bed. 1)

This is the bedroom with the corner window. All the ventilators have been blocked up and the room smells musty. We would advise that these ventilators be unblocked to allow a good airflow to the room. An electric storage heater was noted in the room. This was not tested.

In places, the picture rail was missing and a skirting board was loose. These require remedial works carrying out. Wallpaper has been damaged in several places and the room will therefore require redecoration. (For details see before.)

Front bedroom (bed. 2)

Hair crack noted over window head. This crack does not seem to be of a serious nature. Dimplex heater noted in room. This heater has not been tested. (For details see before.)

Rear bedroom (bed. 3)

A storage heater was noted in this room. This has not been tested.

The cill board to this window was not replaced when the windows were renewed. Some decoration needed around windows to make good where the old frames have been taken away.

The airing cupboard is in this room and contains the water tank and a lagged cylinder. It was not possible to inspect the inside of the tank because of its location. (The top of the tank was against the ceiling.) As rust was apparent on the outer surface, we believe that there is a strong possibility of the tank bursting within the near future and would advise that this tank be renewed as soon as possible. The doors to the cupboard are very stiff and require easing. Picture rail and skirtings missing in one alcove and require replacing. (For details see before.)

EXTERNALLY

Rear gardens

The fence to the left (looking from front of house) comprises concrete gravel board and interwoven fencing and is in reasonable condition. Fence right at bottom of garden is in a similar condition. (Both need a coat of creosote.) Hedge on right needs clipping.

The garden is generally well kept and is fairly large by modern standards and contains an ornamental pond. There is a group of three poplars in the rear garden and also a very large oak tree in the neighbour's garden (number 1). In time, these trees could affect the property (if they are not doing so already).

All the trees are within 20 ft (6 m) of the house. Their presence concerns us.

We have checked the heights and the distances against National House-Builders Council recommendations and believe that the roots of these trees represent a real threat to the house, especially during long dry summers. As the poplars pose the greatest threat, having roots that spread over a very large area, we believe that they should be removed by a competent tree surgeon unless it can be established that the house foundations are in excess of 2.1 m deep (which is unlikely). We would also advise that the tree surgeon should have insurance cover of not less than £. . . . . (Should you require a list of our approved tree surgeons we will be pleased to supply same.)

NB The trees should only be cut down after checks have been made with the Planning Department concerning possible tree preservation orders.

Front garden

The left-hand front gate post is snapped off at ground level and needs replacing. The gate needs rubbing down, re-priming and repainting. The front fence is in need of attention. There are props at various places to prevent the fence from falling over. The fence with number 1 is also in poor condition. The hedge on the right-hand side needs clipping.

The driveway is concrete and has cracked in several locations but this is not unusual with paving of this nature.

The front door step to the rear side door is unfinished and requires covering in quarry tiles.

RATING, PLANNING, BUILDING CONTROL AND COAL BOARD

We have not made any enquiries regarding rating, planning or Building Control approvals for the two extensions or sent off for a Coal Board survey. From your instructions, we were led to believe that your solicitor was making these enquiries.

We would recommend that your solicitor makes enquiries concerning the boundary between numbers 3 and 5. Some past encroachment does appear to have occurred.

SCHEDULE OF COSTS

A schedule of approximate assessed costs is listed below for your reference:

Note
SCHEDULE NOT REPRODUCED.

APPROXIMATE TOTAL   £. . . . .

THESE COSTS DO NOT INCLUDE ANY VALUE FOR ANY WOOD-WORM OR DRY ROT WORK.

CONCLUSIONS

You will note within the report several instances where alteration/remedial work seems to have been carried out in an unworkmanlike manner.

You are strongly recommended not to purchase this property until all the specialist reports and additional inspections have been made.

It is also recommended that the purchase price be re-negotiated to a more realistic figure.

A. R. WILLIAMS MCIOB

For and on behalf of
Building Services (Technical)

Appendix D: Hinter Lane report

For

19 Hinter Lane,
Magpie Hill,
Near Tipham,
Tipfordshire

For

Mr Dove,
95 Slag Heap View,
Tipham,
Tipfordshire

Building Services (Technical)
Old Ferry Offices,
The Square,
Tipham,
Tipfordshire

February, 19—

NOTE
THIS REPORT WAS CARRIED OUT FOLLOWING TIMBER TREATMENTS BY SPECIALISTS. ASSURANCES WERE GIVEN THAT ALL WAS IN ORDER. DURING THE SURVEY FURTHER OUTBREAKS OF WOODROT WERE DISCOVERED.

NOTE ALSO THAT IT WAS DISCOVERED THAT UNDERPINNING WORKS HAD BEEN CARRIED OUT WITHOUT BEING SUBJECTED TO BUILDING CONTROL SUPERVISION.

Although the following report is based upon a genuine survey, the names of any persons, companies, properties and the locations have been deliberately changed. Building Services (Technical) is a registered business name (Reg. no. 2175590).

The various sketches and documents referred to in the text of the main report have not been reproduced.

References to sketches in the report have been dashed out in order that there is no confusion with those referred to in the rest of the book.

The report has not been revised to cover the latest changes to the Insulation Standards, particularly subfloor insulation.

INSPECTION OF 19 HINTER LANE, MAGPIE HILL, NEAR TIPHAM, TIPFORDSHIRE

INSTRUCTIONS, LIMITATIONS AND GENERAL PREAMBLE

We report that we carried out a survey at the above property on Monday —th February, 19— as instructed in accordance with our standard conditions of engagement (as previously forwarded, see copy enclosed).

This report shall be for the private and confidential use of our client for whom the report is undertaken and must not be reproduced in whole or in part or relied upon by a third party.

We will not be responsible to third parties who obtain, by any means, a copy of this report and act upon information contained therein.

The survey commenced at approximately 1 p.m.

In the morning it had been raining heavily and there were high winds but during the course of the survey the conditions improved.

As you are aware, the present owner has had remedial works carried out by Tipham Damp Proofing (Often Narrows) Ltd (from now on referred to as the treatment company).

At the end of this document, we have (with permission) bound in several reports from the treatment company which provide additional details of the remedial works that they have carried out. (*NB* We have generally inspected works in areas that are easily accessible, and as far as can be ascertained, all seems to be in order but we cannot give any guarantees as to the effectiveness of any of the systems used. In the event of any future outbreaks of wood infestation, woodrot or rising dampness, you must look to the guarantees provided by the treatment company.)

The works that will be guaranteed by the treatment company (subject to specified additional builders' work being carried out) are as follows:

(a) The injection damp courses which prevent rising dampness. The treatment company have stated that they have injected all walls in the property with the exception of the internal walls surrounding the bathroom/WC/utility area.

(b) The ground floor is guaranteed against future woodworm infestation. The roof space has not been treated.

(c) Joists in the areas listed in the reports from the treatment company have been replaced on the ground floor. The roof space, according to treatment company, is free from defect from infestation or fungal decay.

It is worth noting however that the treatment company require plaster-work renewing using their patent system for the first metre of wall. This must be done carefully or their guarantee will be void.

In addition, the treatment company have advised in their letter of —th March, 19— that other remedial works must be put in hand as part of a separate contract (which is to be carried out by others) and that their guarantees will not become effective until the listed works are put in hand. These works are as follows:

(a) Overhauling the roof covering
(b) Overhauling gutters and downpipes
(c) Repointing all external walls
(d) Provision of adequate ventilation to the subfloor voids and voids in the roof space.

We would report that *all* of these items still require attention.

It is also worth noting that if new floorboards are used anywhere, they require spraying with approved fluids to comply with the treatment company recommendations.

Whilst every care has been taken in completing our survey and report, the investigations have been non-destructive in nature and therefore we are unable to report upon matters which were concealed at the time of the inspection and the following assumptions have had to be made:

(a) That high alumina cement (HAC), concrete or calcium chloride additive or other deleterious materials or techniques were not used when the original property or any subsequent extensions were built.
(b) That any wall ties that exist are not perished.
(c) That your solicitor's or legal adviser's search will prove that the property is not subject to any unusual covenants, Local Authority restrictions, or onerous restrictions imposed by others, encumbrances or outgoings and that good title can be shown.
(d) That future inspections of any hidden parts of the structure which have not been inspected during this survey will not reveal material defects, or cause the surveyor to alter materially the indicative costings shown at the end of this report (where applicable).
(e) That the treatment company has diligently carried out its remedial works and that it has efficiently remedied all outbreaks of woodrot, timber infestation and rising dampness.

As you are aware, it was possible to gain access beneath most of the suspended timber ground floors and the open areas of the roof span but there are obviously areas which could not be uncovered, exposed or made accessible and we are therefore unable to report that the property is free from defect; however we have done our best to draw conclusions about the construction from surface evidence visible at the time of our inspection. We have not inspected any cavities in walls or the condition of any wall ties (if applicable). As indicated later, we are of the opinion that the house walls are of solid construction.

References to left and right are made facing the property standing in Hinter Lane.

THE PROPERTY GENERALLY

The property is a detached bungalow. It is a traditional brick built structure with a slate roof. The ground floor is suspended timber construction.

From discussions with the present owner we understand that the bungalow was built around 1933 and has been unoccupied for about 18 months. We have not passed any comments on the two rear 'outriggers' as we have been instructed to ignore same. (We understand that it is your intention to demolish same should you proceed to purchase.)

The walls are constructed of red stock brick on the front elevation and common brickwork to the sides and back. Judging by the header courses in the walls, the external brickwork would seem to be 9 in (225 mm) solid walling. The building was of simple construction having two gable ends. No test was made on wall ties (not applicable to solid walls anyway) as this would require drilling the walls or cutting out brickwork and inspecting the cavities either visually or with an endoscope, which is beyond the terms of our engagement.

The property comprised of a porch, hall, bathroom, scullery, larder, kitchen, lounge and two bedrooms. Externally there was a tumbledown greenhouse and a garage.

As you are familiar with the property and area, we have not provided superfluous information such as room sizes, the location, schools in the area, etc.

THE MAIN STRUCTURE

The main pitched roof

Where practicable, the interior of the main pitched roof has been examined but the inspection of the roof covering has been made from ground and eaves level only.

The main roof is a traditional rafter and purlin construction covered with slates. The roof has no sarking felt beneath. (Sarking felt is a secondary layer normally incorporated into new properties to prevent water penetration if a slate is dislodged.) The slate roof seemed sound when visually inspected from the outside (with the exception of two slipped slates noted on the rear elevation) but the lack of sarking felt and the moisture levels inside the loft space incline us to advise that the slates be carefully removed, the roof re-battened and felted and the slates replaced.

The ridge tiles and verges seemed to require repointing. If the roof were to be relaid, then obviously these defects would be remedied.

We gained access to the loft through a large glazed trapdoor/borrowed light in the hall. The trapdoor was well sited and of reasonable size. We would suggest that the glazing be replaced with toughened glass or a plywood panel, as injury could result if there were an accident.

The loft was filthy. Most of the timbers to the loft floor were covered in accumulated soot deposits and it was impossible to fully check the ceiling joists. We tested the timbers in several locations with a moisture meter.

In all areas the readings were well above normal (i.e. if the timbers are not dried out in the near future, dry or wet rot outbreaks could be anticipated if they have not already started).

As the house has had rot outbreaks in the floor, there is a strong possibility that some outbreaks could be anticipated in the roof (despite the assurances from the treatment company).

This is the main reason for our recommendation to re-roof. A re-roofing operation would ensure that water cannot enter the loft and also any timbers likely to be rotten can be located, cut out and treated.

Around the two chimneys the timbers had exceptionally high moisture readings. The brickwork to the chimeys also had very high moisture levels. Our inspection revealed that several flashings were loose. (Flashings and soaks are purpose-made pieces of lead or lead substitute inserted to fill in a junction between the chimney and roof.)

We would suggest that the flashings and soakers are defective and should be replaced in code 4 and 5 lead.

The ceiling is uninsulated. This is totally inadequate. Under modern regulations the insulation layer should be at least 100 mm thick. (*NB* This thickness could soon be increased to 150 mm). The chimney breasts in the roof spaces require repointing and rendering.

The pipe lagging was 'do-it-yourself' and comprised of old papers cut into strips and attached to the pipes with hessian.

Where exposed, the pipe work was noted as being of lead. We would recommend that the old lead pipes be replaced with copper and insulated properly with patent lagging. One copper pipe was noted in the roof space but judging by its location it was a gas feed pipe.

There is an old water tank in the loft which is redundant. This should be removed. The water tank (new plastic) seemed in good condition but as the system had been drained down it was not possible to test. This tank requires lagging. From visual inspection, the ceilings appear to be plasterboard.

Roofs to bay windows (3 no.)

Above the bays the lead coverings are perished and the flashings are loose. The front left-hand bay has evidence of patching with felt. We would advise that all leadwork to the bay roofs be replaced. The flashings above the bay aprons were loose. These should be refixed and pointed. As the ceilings to the bay have suffered because of the leaks, these require replacement.

Chimneys (2 no.)

There were two chimneys. Both require being fully repointed. It would also be advisable to re-flaunch the tops of the chimneys. (Flaunching is the cement and sand topping to the chimney.) The lead flashings also require attention as previously mentioned. At the time of inspection all the fireplaces were blocked up. Judging by the soot falls discovered the flues require sweeping.

Main walls of house

No examination has been made of the foundations because this would require excavating trial pits. However, from internal inspection below the ground floor, it would appear that the bungalow has been built on spread brick foundations (i.e. not concrete strip footings). Evidence was found of some concrete footings and, whilst it is not possible to be absolutely positive, it does seem that at some date after initial construction that the front and back walls were underpinned (i.e. new foundations inserted under the old ones).

There was evidence of a damp proof course (DPC) in the external walls. We also noted bore holes for the new injection DPC. We have inspected the treatment company's standard guarantee dated October 19— which would appear to guarantee the work for 30 years. (NB This guarantee does not become effective until plastering works have been completed.)

As this company has a good reputation locally and has its guarantees backed by the manufacturer of the treatment fluids, all would seem in order. (Note later in the report that a further outbreak of wood rot was discovered and this must be reported to the treatment company for action under their timber treatment guarantee.)

The bay windows (3 no.) appear to becoming detached from the main house. The right-hand rear bay is particularly bad and will require demolition and rebuilding. The front right-hand bay will also require some attention. Below DPC level generally there were signs of brick spalling in several places (i.e. bricks breaking up owing to frost action). These bricks should be cut out and replaced.

A repoint is required to both gable ends and possibly to the front and back elevations. This should be carried out by raking out the joints to a minimum of 12 mm and flushing up with a suitable mortar mix. The rear main wall to the house will need rebuilding in places.

Drainage

There would appear to be two manholes in the garden. Neither had suitable covers (merely flagstones laid across them). These temporary covers require replacing with purpose-made steel covers.

When water was run, it was obvious that these drains were not the ones serving the foul system. We are therefore unable to comment upon the drains except to say that the pipes were virtually on the surface of the ground. A spigot of one pipe was noted as being broken. We would suggest from the damage noted and the cluster of small shrubs close to the drain that this pipework and the manholes could need replacement.

Windows, doors and timberwork

The windows were boarded up, so access was restricted.

The external doors, windows, fascias and other boarding were in poor condition and will need extensive repairs/renewal.

The moisture meter indicated high readings generally, and when the prongs were inserted, the timber in areas of high exposure were found to be spongy (e.g. feet of door frames, window boards, external cills). The bay windows had also started to spread as a result of movement. We would envisage that the windows will need replacing completely, if not immediately, certainly in the not too distant future.

The fascia boards were tested in several areas and seemed in reasonable condition but require 'burning off', repriming and two good coats of paint applying. There probably will be one or two sections that might need renewal.

Gutters and downpipes

The gutters and downpipes had in the most part been replaced by a new PVC system but in the areas inspected appeared to be laid to incorrect falls.

Water was observed standing in all the gutters and several downpipes appeared blocked with leaves and tree litter. We would advise that the gutters and downpipes be thoroughly cleaned out and relaid to steeper falls. At the rear of the property the old cast iron soil stack was pulling away from the wall and will undoubtedly collapse in the near future if not resecured.

INTERIOR GENERALLY

Electric/gas meters

The electric meter and gas meter were inside an old outbuilding at the rear right-hand side of the property. Our brief indicates that these services will be resited if the property is purchased and the outbuildings demolished; therefore comments in this area are somewhat superfluous except that we noted that bricks beneath the steps to the side entrance were dislodged and need rebuilding. (The top three to four courses of the steps to the side door were also in a dangerous condition and need rebuilding.)

Central heating installation

No central heating system was apparent. The only heating in the property appeared to be independent gas fires (which were not tested).

Electrical installation

The electrical system was visually inspected. Our brief indicates that it is your intention to rewire the house if purchased. As most rooms are substandard in having only one or two points per room at the most and seemed to be wired in rubber insulated cables, this would be advisable.

Floors

As previously indicated, we crawled beneath most of the suspended floor areas. The floor space was damp and did not seem to have a subfloor. Moisture readings were high. In some areas (e.g. front and back walls), it appeared that underpinning work had been carried out in the past. We contacted the Local Authority Building Control Department and spoke to their records section. We were informed that their records only went back to 19— but that during the 19— to 19— period they have no record of underpinning work taking place.

There was a large amount of evidence of the treatment company's work (i.e. new joists and floorboards). The work seems reasonably well carried out with each new joist being wrapped in DPC where it came into contact with damp walls. More work below the floors will be necessary, however. In one location bricks were displaced in the sleeper walls (the walls below the floor that support it).

The underfloor areas were dirty, being filled with brick rubble and old pieces of timber. In places, pieces of rotten timber were found lying on the earth. The timber in particular is a breeding ground for wood rot. There was a large amount of efflorescence noticeable in the below ground brickwork. (Efflorescence is a white powder formed when soluble salts are left behind after evaporation of rising water.)

The underfloor ventilation seemed good but one or two airbricks require cleaning out. (The treatment company should be consulted regarding additional airbricks as it is a requirement that additional vents are provided.)

The kitchen floor was solid being covered with red quarry tiles. It was possible to see the full construction of the floor at an intersection with the pantry area.

As the floor seemed to be the old form of construction and would have no damp proof membrane, we would advise that it be dug up and replaced with a new concrete floor comprising 150 mm (6 in) of clean hardcore, visqueen DPM, 100 mm concrete and finished with a 50 mm cement and sand screed and new floor finish provided. It is essential that the new visqueen DPM be taken up the walls and connected to the DPC in the walls otherwise dampness will still affect the kitchen area. During the course of the inspection we noted that the kitchen floor has settled adjacent to the linking door to the hall. When the kitchen floor is relayed, then it should be laid to line up with the timber floors.

Ceilings

The ceilings appeared to be plasterboard. Although cracking was apparent, the pattern of the cracks lead us to conclude that they were caused by expansion and contraction where individual boards met, which is not uncommon.

Wall plaster

In areas the wall plaster was uneven. This was particularly so in the kitchen wall adjacent to the side door. We would advise that this wall have the plaster removed and be replastered using renovation plaster.

Decoration

The property generally was in a poor state of decoration.

GROUND FLOOR

Porch

(For floors, decoration, electrics, etc., see previous comments.)

The porch was an integral construction between the two front bays. The floor of the porch is a mosaic. This floor is likely to be of similar construction to the kitchen. As it will undoubtedly 'sweat' (i.e. let water up to the upper surface), we would advise that the mosaic floor be lifted, the composite parts saved and a new floor laid (specification as for the kitchen) and the mosaic floor relaid.

Hall

(For floors, decoration, electrics, etc., see previous comments.)

The hall ceiling had paper curling off it and when tested had a high moisture reading. This would indicate that the leadwork at the intersection of the porch and main roof required replacing.

Kitchen

(For floors, decoration, electrics, etc., see previous comments.)

An outbreak of wet rot was noted in the skirting by the sink. After studying the fungus pattern and charring effect on the reverse side, we came to the conclusion that it was an outbreak of wet rot (*Coniophora puteana*). This type of rot is not as serious as dry rot but needs treating by Tipham Damp Ltd and, in our opinion, should have been treated during the period when remedial works were carried out.

Other areas of rot were also noted in the woodwork around the sink. We were unable to indentify the type of outbreak but once again would recommend immediate treatment in case they turn out to be true dry rot which can spread very quickly.

WC, bathroom and pantry

(For floors, decoration, electric, etc., see previous comments.)

These rooms were extremely small and unsuitable for a modern house. We would envisage that the walls between the WC/Pantry and bath area would require demolishing and combining into one room. As they are of similar size and condition we have considered them together.

The so-called bathroom and WC areas only have the minimum of facilities. There was no bath apparent at the time of inspection, and if there ever had been, it would be difficult to envisage where it would have been sited.

Rot was apparent on the windows and window boards. Lead piping served the fittings. This should be stripped out and replaced in copper. A leak was noted under the wash hand basin.

Three main rooms

(For floors, decoration, electric, etc., see previous comments.)

These rooms were of similar size and condition and have therefore been considered together. The ceilings above the chimneys were damp, once again indicating leaks from the flashings and the fact that the fires had not been in use for some time. We believe that water has entered the flues of the chimneys and is now causing damp internally. As previously indicated, the bays are in poor condition. The ceilings of all the bays are suffering from water penetration from the defective lead roofs above and require hacking down and replacing. In the front right-hand room (lounge) there was a copper cylinder in the cupboard by the fireplace. The size seems somewhat inadequate. It was also uninsulated. We would recommend that a properly lagged larger cylinder with an integral immersion heater be installed. Some areas of plaster in the cupboard were also missing.

The rear fireplace had tiles missing which require replacement.

EXTERNALLY

Generally

The gardens are overgrown and very damp. The paths need to be cleaned down and relevelled. The garden is surrounded on all sides by old stone walls. These need pointing on all sides. Where necessary the luxuriant growths of ivy need cutting back to facilitate this operation.

Rear garden

Growths of saplings were noted near the rear wall of the house. These should be cut down and poisoned because if allowed to flourish they will affect the stability of the walls. The rear garden is stepped in several places. The walls at the steppings are in bad condition and require rebuilding. Rubbish in the garden needs carting off site.

Front garden

Generally overgrown but in better condition than the back.

Left-hand area

This is overgrown and the grass requires cutting. The fence between the side area and the rear has most of its panels missing.

Right-hand garden area

The right-hand garden area was overgrown.

This area contains a garage and old greenhouse. Both are in poor condition. The greenhouse, for all intents and purposes, may as well be demolished as it no longer contains any glass.

The garage will require a new roof as the felt has perished and most of the roof boarding will undoubtedly be affected by wood rot. The doors and windows are badly rotten and will require replacing. Inside the garage there is a pit. True dry rot (*Serpula lacrymans*) noted on the timbers covering the pit. These timbers should be destroyed and the pit cleaned out as the rubbish can only harbour diseases which could infect the house.

RATING, PLANNING, BUILDING CONTROL

We spoke to the Planning Department. The Council official that we contacted informed us that outline approval has been granted on the right-hand garden area for a two storey dwelling and garage. (Application number . . . . . granted —rd February, 19—.)

As previously mentioned, we made enquiries with Building Control concerning the underpinning but they have no records of this work. We have not made any enquiries regarding rates or Coal Board searches and have presumed that your solicitor will make these enquiries.

CONCLUSION

The house is in poor condition despite the work already carried out by the treatment company. As we have inspected 30 year guarantees, we would conclude that there is little to fear regarding this part of the project. However, we believe that further work will be required.

As promised, we have included below approximate costings to cover likely remedial work. However, we would stress that these are only very approximate and are intended to be indicative and not an exact cost.

A. R. WILLIAMS MCIOB

For and on behalf of
Building Services (Technical)

Appendix E: Fern Lea report

For

Fern Lea,
5 Slag Heap View,
Tipham,
Tipfordshire

For

Mr Brown,
Rugby House,
Odd Balls Lane,
Tipham,
Tipfordshire

Building Services (Technical)
Old Ferry Offices,
The Square,
Tipham,
Tipfordshire

July, 19—

NOTE
THIS REPORT HAS BEEN INCLUDED TO ILLUSTRATE THE EFFECTS OF A SULPHATE ATTACK. PRIOR TO COMMENCEMENT OF THE SURVEY, THE OWNERS OF THE PROPERTY CLAIMED THAT THE HOUSE HAD ONLY JUST BEEN SURVEYED AND THAT IT HAD BEEN FOUND FREE FROM DEFECT. THESE CLAIMS WHICH WERE OBVIOUSLY INTENDED TO MISLEAD WERE PROVED FALSE.

Although the following report is based upon a genuine survey, the names of any persons, companies, properties and the locations have been deliberately changed. Building Services (Technical) is a registered business name (Reg. no. 2175590).

The various sketches and documents referred to in the text of the main report have not been reproduced.

References to sketches in the report have been dashed out in order that there is no confusion with those referred to in the rest of the book.

The report has not been revised to allow for recent changes in the Building Regulations. Note in particular that recommended thicknesses of insulation have been changed since the original report was written.

DOMESTIC BUILDING REPORT ON FERN LEA, 5 SLAG HEAP VIEW, TIPHAM, TIPFORDSHIRE

INSTRUCTIONS, LIMITATIONS AND GENERAL PREAMBLE

We report that we carried out a survey at the above propery on —th July, 19— as instructed, in accordance with our standard conditions of engagement (as previously forwarded, copy bound into rear of this report).

This report shall be for the private and confidential use of our client for whom the report is undertaken and must not be reproduced in whole or in part or relied upon by a third party.

We will not be responsible to third parties who obtain, by any means, a copy of this report and act upon information contained therein.

Whilst every care has been taken in completing the survey, the investigations have been non-destructive in nature and therefore we are unable to report upon matters which were concealed at the time of the inspection. We have not inspected woodwork or other parts of the structure which are not uncovered, unexposed or are inaccessible and we are therefore unable to report that the property is free from defect; however, we have done our best to draw conclusions about the construction from surface evidence visible at the time of the inspection. We have not inspected the wall cavities or wall ties.

It was not possible to fully examine the floors because there were fitted carpets or permanent coverings in all rooms. Corners of carpets were lifted where possible and isolated areas examined. In several areas, it was obvious that the floors appeared to be under extensive sulphate attack. However, in order to be one hundred per cent certain of the cause of the floor failure laboratory tests would be required which are beyond the scope of this report. This aspect is discussed in greater detail later.

The following assumptions have had to be made when preparing this report:

(a) That high alumina cement (HAC), concrete or calcium chloride additive or other deleterious materials or techniques was not used when the property or any extensions were built.
(b) That any wall ties that exist are not perished.
(c) That your solicitor's search will prove that the property is not subject

to any unusual or especially onerous restrictions, encumbrances or outgoings and that good title can be shown.

(d) That inspection of those parts which have not been inspected would neither reveal material defects nor cause the surveyor to alter materially any conclusions.

The survey commenced at approximately 1 p.m. and it was raining heavily.

References to left and right are made facing the property standing in Slag Heap View.

THE PROPERTY GENERALLY

The property is a semi-detached bungalow. From discussions with the present owner, we understand that the original property was built in 1931/2. It has subsequently been extended at the rear.

The building is a traditional brick built structure with a red tile roof. One side of the roof is hipped and the hips are covered with purpose-made tiles. The external walls comprise reasonable quality rustic facing bricks.

Judging by the stretcher bonding of the brickwork and the general construction, the external walls appeared to be of cavity construction. No test has been made on the wall ties as this would require drilling the walls and inspecting the cavities with an endoscope which is beyond the terms of our engagement.

The property comprised a porch, hall, bathroom, living room, utility room, kitchen, dining room and bedroom.

As you are familiar with the property and area, we have not provided superfluous information such as the location, schools in the area, etc.

THE MAIN STRUCTURE

The main pitched roof

The main roof is of traditional rafter and purlin construction and was covered in red tiles. It also has sarking felt beneath. (This is a secondary layer normally incorporated into new properties to prevent water penetration if a tile is dislodged.) The tiled roof seemed sound when inspected from the outside. A tile is broken on the rear elevation and several cracked tiles were noted on the side elevation. There were areas of the roof which seemed 'wavy'. When we placed a level on the ceiling joists it was noted that there appeared to be a slight slope, which would indicate that some settlement has taken place, or that the joists were not laid true when the property was originally built.

Where practicable the interior of the main pitched roof has been examined but the inspection of the roof covering has been made from ground level only. We gained access to the loft through a trapdoor in the hall. The trapdoor was well sited and of reasonable size.

The loft generally was clean. We tested the timbers in several locations with a moisture meter and generally speaking found the readings were within acceptable limits. The main exception to this was around the chimney stacks. The structural members around the chimneys and gutter boards showed high moisture readings which would lead us to believe that the flashings and soakers require replacement. (These are purpose-made pieces of lead inserted to fill in a junction between the chimney and the roof.)

There are three piers in the loft space which support the purlins (a purlin is a main timber ring beam which holds up the roof). One of the piers had several loose bricks and these need remortaring into position. By inspection, the piers seem inadequate and the loose bricks appear to confirm that they need stiffening as they must be under stress. These should be repaired by a competent bricklayer and enlarged as necessary, taking support from the brick wall below and not from any roof timbers.

The main water tank appeared to be galvanized steel but, because of its location, it was not possible to fully inspect it. However, we have no reason to believe that it was in bad condition. The header tank to the boiler was PVC and seemed in good condition. The main tank was lagged but the header tank was not and we would advise that this be done. All pipework in the loft seemed well lagged.

The main roof was insulated with approximately 3 in (75 mm) of fibre-glass. The normal accepted standard for roof insulation is now 4 in (100 mm).

Two flat roofs to new extensions

The new kitchen extension and the utility room extensions are covered with mineral felt finished in limestone chippings. Both roofs are covered with moss which would indicate that the roofs do not drain well. Where the new flat roofs abut the tiled main roof, there are several areas of irregular tiling where the new work has disrupted the older construction. Although no signs could be found of leaks internally, it is worth noting that the standard BS 747 felt roofs, which these are, have a limited life (perhaps ten years), and if the roofs are not leaking at the moment, there can be no doubt that at some stage in the near future remedial works to the flat roofs will be necessary.

Chimneys (no. 2)

There are two chimneys. Both require repointing especially the top four or five courses. It would also be advisable to re-flaunch the tops of the chimneys. (Flaunching is the cement and sand topping to the chimney.) The lead flashings also require attention.

Main walls of house

No examination has been made of the foundations because this would require excavating inspection pits (trial holes). As the floors seem to be subject to sulphate attack (see later), we would advise that the footings have trial holes dug down and an inspection made of the concrete foundation whilst the floors are being replaced, in case the concrete in the foundations has also suffered. If in doubt a structural engineer should be consulted.

There was evidence of a damp proof course (DPC) in the external walls. To the rear and sides, the DPC is generally 150 mm (6 in) above ground level and about 300 mm (12 in) at the front. These locations are satisfactory and comply with the Building Regulations. However, tests internally indicate that the DPC is not totally effective. We would advise that this be inspected by a firm of injection specialists prior to purchase.

A crack was noted under the window to the dining room. This should be raked out and re-filled. The front bay below DPC level had a slight bulge in it. Adjacent to the porch, there are other areas which showed signs of disruption or spalling. (Spalling is where the brick starts to crumble owing to frost action.) All spalled bricks below DPC level should be cut out and replaced. The brickwork below DPC level was covered in moss. This would indicate that the walls below DPC are very damp. The mortar generally below DPC level was in poor condition. We would recommend that all brickwork below DPC level be raked out and repointed.

When we checked the walls with a large spirit level, we found that the walls in the main were plumb. However, below DPC the walls were generally not as true as the rest of the structure. (See later for details of sulphate attack.)

Drainage

We lifted the cover off the manhole adjacent to the porch and ran the sanitary fittings internally. From visual inspection the water seemed to flow freely and be in working order. No pressure, smoke or other tests have been applied to the drains.

Windows, doors and timberwork

The older windows and doors are generally in better and condition than the ones installed in the two extensions. These doors and windows should be burnt off and repainted.

Gutters and downpipes

The gutters to the older part of the property appeared to be timber. As the adjoining property has had all its gutters replaced in PVC and as extensive leaks were noted on all the wooden gutters, it seems fair to assume that these will need replacing. In the recess between the two rear extensions, there was moss growing on the wall under the gutter over the bathroom. This would indicate that the gutter at this point has been leaking for a considerable amount of time.

The new porch to the side has been built around the old timber gutter. When this gutter was tested internally, it was found to be rotten. In order to replace this gutter, it will be necessary to replace the porch roof also.

The older part of the building has cast iron downpipes. These were all rusted at the backs and have not been painted where hidden from view. As a swan-neck (pipe comprising two bends with a short length of pipe between to connect the main pipework to the overhanging gutters) has been replaced near the porch, we would expect that the rest of the downpipes will need replacing in the near future. The extensions generally had PVC gutters and downpipes which seemed in reasonable condition.

INTERIOR GENERALLY

Gas meter/gas supply

The gas meter was in a cupboard in the hall. No test was made to the gas supply, but as the heating system was functioning at the time of survey, we have no reason to doubt that the system functions adequately.

Central heating installation

The central heating installation was powered by a Glow-worm Space-saver 38 which has a balanced flue outlet in the kitchen. The system has not been tested, but from visual inspection it would appear to be in good order as the house was warm and all the radiators appeared to work.

There appeared to be radiators in all rooms except the kitchen. All pipes seemed to be surface mounted and, where running down walls, covered by plastic trunking which was rather unsightly.

Electrical installation

The electric meter was in the hall and the consumer unit was controlled by a six-way fuse board. We were unable to ascertain whether the system was a modern PVC installation. Prior to purchase, I would advise that an approved electrician be asked to check the system as it may prove that there are rubber insulated cables in existence which tend to perish after 25 years' life. Any rubber insulated cable should be replaced by PVC covered cable. The final system should comply with the IEE regulations.

The points noted were as follows:

	Light points	Sockets	Other
Porch	—	—	—
Bathroom	1		cooker
Kitchen	2	4 doubles	
Dining room	1	1 single	—
Hall	1		
Bedroom	1	1 single	
Lounge	1	2 single	
		1 double	
Ironing room			
Box room	1	1 double	

Floors

As previously indicated, we are of the opinion that the solid ground floors are being subjected to what is known as a sulphate attack. We would stress that we were unable to lift all the floor coverings, but from what we have managed to uncover, we believe that this defect affects all the older floors in the house and it is also possible that it could be affecting the newer floors.

Sulphate attack is a process whereby concrete crumbles because sulphates in solution start to attack it. The usual symptoms are 'mole hills' forming in the floors, and extensive cracking.

Surrounding brickwork can also be affected by the 'heave'. We suspect that the cause of the problem is that ash has been used to 'blind' the hardcore bed beneath the floor. The use of ash was fairly common in the area because it was readily available.

(Blinding is a thin layer of soft material that covers the top of the hardcore fill below the floor.) We believe that the high water table has carried the sulphates contained within the ash through to the concrete in the floor slab. A chemical reaction has then taken place over a long period of time which is causing the concrete to expand and degrade.

The only remedy that we know of to deal with this problem is to break up the existing floors, dig down and remove the old hardcore under and relay the hardcore and blinding with new uncontaminated material. This is not a defect to ignore. We would advise that several reputable builders be asked to provide quotations for this work as we would envisage that the work would be very costly.

We discovered cracks in the dining area and hall floors ranging from 6 mm (¼ in) to 13 mm (½ in) in width. Some attempt had been made to fill these cracks with patent filler, but as the cracks have widened this filler has dropped out. We therefore conclude that the attack (which is usually a slow process) is still underway and the situation will only get worse as the years proceed.

Decoration

The property generally is in a reasonable state of decoration but no doubt will suffer damage if extensive building work replacing the floor slabs is not put in hand.

GROUND FLOOR

Porch

The porch appeared to have been built on what is known as a 'raft'. For lightly loaded structures, this form of construction is usually adequate.

We placed a long spirit level on the quarry tiled floor of the porch and found that it seemed out of true. The interior of the porch was not plastered.

(Note also our previous comments on the porch roof.)

Utility room

The utility room is very narrow. Where the new floor joins the older floor there is an irregularity.

Bathroom

We tested the fittings by running water and all seemed in order. We did not test the electric shower. The bath was made of good quality acrylic and had a shower attachment. The WC and wash basin seemed in order. The floor to the bathroom seemed to have been replaced but we would advise that this be checked by a builder prior to purchase, bearing in mind our previous comment on the floors generally. The floor when tested seemed out of level. The bathroom was half tiled. The tiles are black and not very attractive. Four tiles were damaged over the wash basins. Above the tiles the walls are finished with Polytex or similar do-it-yourself Artex substitute, which was very rough.

We always suspect areas where this kind of finish has been applied because we find that this type of product is used to disguise defects in plasterwork. You must anticipate having to carry out some remedial works to the bathroom walls.

Kitchen

The kitchen units are of reasonable quality but the Moffat hob is well used and needs replacing.

The kitchen is contained within one of the new extensions and connects with the dining room. As there is no door between the two rooms, smells will obviously travel. The kitchen floor was out of level and, as a consequence, all the doors to the units are misaligned.

Dining room

As previously mentioned, the dining room floor has suffered from what appears to be an extensive sulphate attack and there seemed to be damp in the outer walls. We suspect that the defective floors have caused heave on the outer walls and have affected the DPC.

Hall

As previously detailed, there appears to be a sulphate attack on the floor. The hall had a gas radiator as well as a radiator off the main system. The gas radiator was not tested.

Bedroom

There were high moisture readings in the walls.

Lounge

High moisture readings in the walls.

EXTERNALLY

Rear gardens

The garden is generally well kept but very small. The left-hand wall is in reasonable condition but the render is slightly crazed in places. This may need attention in the future.

The right-hand fence and rear fence are in poor condition and need several replacement interwoven panels inserting. Where inspected (there are large shrubs growing along the boundary and some parts of the fence were not visible), at least four panels need replacing on the left-hand fence. The patio was composed of 600 × 600 red and grey concrete flags and seemed in good condition.

There is a small timber shed in the garden. This seemed in good condition, but as it was locked, we did not inspect it internally.

Side garden

There is a narrow strip comprising a concrete path and a border which is heavily planted with climbing plants such as ivy and shrubs.

The concrete path is damaged in one area adjacent to the porch and requires filling.

Front garden

On the left-hand side there is a low curved wall. The curved section has several upper courses collapsing. These require rebuilding. The external face has what appears to be a heavy moss covering and needs cleaning down.

The front wall requires repointing front and back. The coping bricks are badly spalled. The spalled bricks should be cut out and replaced as necessary. The front path by the bay has settled and requires relaying.

RATING, PLANNING, BUILDING CONTROL AND COAL BOARD

We have not made any enquiries regarding rating, planning or Building Control approvals for the two extensions and have presumed that your solicitors will make these enquiries. We did speak to a local Building Control Officer concerning ground conditions in the area and were told that the substrata in this area is generally rock which, of course, is a good bearing course for the house. We will forward the Coal Board survey directly upon receipt.

CONCLUSIONS

The property is in a good area and, according to Building Control, the house is founded on rock.

However, this house requires a great deal of attention. If enough money is spent on it, it could have potential. We would stress that we would envisage that the cost of rectification would cost many thousands of pounds and would suggest that the owner be asked if he intends to rectify this work himself. He did suggest that he would be approaching his insurers.

In our opinion, it would be an unwise move to purchase the property unless some idea of likely cost of rectification were obtained prior to purchase or that the defects be rectified in total by the present owner.

A. R. WILLIAMS MCIOB

For and on behalf of
Building Services (Technical)

Appendix F: Snow Road report

For

6 Snow Road,
Tipham,
Tipfordshire

For

Mr Brown,
Rugby House,
Odd Balls Lane,
Tipham,
Tipfordshire

Building Services (Technical)
Old Ferry Offices,
The Square,
Tipham,
Tipfordshire

October, 19—

NOTE
THIS REPORT HAS BEEN INCLUDED TO COUNTERBALANCE
THE OTHERS. NOT ALL HOUSES HAVE MAJOR PROBLEMS!

Although the following report is based upon a genuine survey, the names of any persons, companies, properties and the locations have been deliberately changed. Building Services (Technical) is a registered business name (Reg. no. 2175590).

The Coal Board report, and the various sketches and documents referred to in the text of the main report, have not been reproduced.

References to sketches in the report have been dashed out in order that there is no confusion with those referred to in the rest of the book.

The report has not been revised to allow for recent changes in the Building Regulations. Note in particular that recommended thicknesses of insulation have been changed since the original report was written.

Note the clause inserted to cover cavity fill in walls.

BUILDING SURVEY ON 6 SNOW ROAD, TIPHAM, TIPFORDSHIRE

INSTRUCTIONS, LIMITATIONS AND GENERAL PREAMBLE

We report that we carried out a survey at the above property on Sunday, —nd October — as instructed, in accordance with our standard conditions of engagement (as previously forwarded, see copy bound into rear of this report).

This report shall be for the private and confidential use of our client for whom the report is undertaken and must not be reproduced in whole or in part or relied upon by a third party.

We will not be responsible to third parties who obtain, by any means, a copy of this report and act upon information contained therein.

We have not inspected woodwork or other parts of the structure which are not uncovered, unexposed or are inaccessible and we are therefore unable to report that the property is free from defect; however, we have done our best to draw conclusions about the construction from surface evidence visible at the time of our inspection. We have not inspected the wall cavities, cavity insulation or wall ties.

Whilst every care has been taken in completing our survey and report, the investigations have been non-destructive in nature and therefore we are unable to report upon matters which were concealed at the time of the inspection and the following assumptions have had to be made:

(a) That high alumina cement (HAC), concrete or calcium chloride additive or other deleterious materials or techniques were not used when the original property or any subsequent extensions were built.

(b) That any wall ties that exist are not perished.

(c) That your solicitor's or legal adviser's search will prove that the property is not subject to any unusual covenants, Local Authority restrictions, or onerous restrictions imposed by others, encumbrances or outgoings and that good title can be shown.

(d) That future inspections of any hidden parts of the structure which have not been inspected during this survey will not reveal material defects, or cause the surveyor to alter materially the indicative costings shown at the end of this report (if applicable).

It was not possible to fully examine the floors because there were fitted carpets or permanent coverings in all rooms. Corners of carpets were lifted where possible and isolated areas examined.

No pressure, smoke or other tests have been applied to the drains.

The survey commenced at approximately 10.00 a.m. The weather conditions were fine but it had rained heavily the night before.

References to left and right are made facing the property standing in Snow Road (i.e. the left adjoins number 8).

SPECIFIC INSTRUCTIONS

We were particularly asked to inspect the roof to see if it would be possible to carry out a loft conversion.

THE PROPERTY GENERALLY

The property is a detached true bungalow, which according to the present owner is about 20 years old.

From observations made at the bungalow, the rear garden gets the sun in the morning.

The building is a traditional brick built structure (with feature tiling to one panel on the front elevation) and a tiled roof.

The accommodation internally comprised a hall, rear living room, kitchen, two bedrooms, bathroom, airing cupboard and cloak cupboard. Externally there is a detached garage which is served by a flagged drive some 8 ft 6 in (3 m) wide. (This width is better than most small houses built today). It is worth noting, however, that the eaves of number 4 project over the drive of number 6 (as 6 projects over 8).

This is quite normal as long as the deeds of the property cover the point (a word with your solicitor might be advisable), but in the event of one wishing to build over the drive, the projection of the adjoining property would obviously hinder this. It was also noted that number 8 has a carport which is attached to the overhang of number 6. (Presumably with the present owners' agreement.)

As you are familiar with the property and area, we have not provided superfluous information such as the room sizes, location, schools in the area, etc.

For reference, we have bound into the rear of the report photocopies of the survey sketch and photographs for your perusal.

We would reiterate that no test has been made on the wall ties as this would require drilling the walls and inspecting the cavities with an endoscope which is beyond the terms of our engagement.

THE MAIN STRUCTURE

The main pitched roof

Where practicable, the interior of the main pitched roof has been examined but the inspection of the roof covering has been made from ground and eaves level only.

The roof was of traditional rafter and purlin construction (a purlin is a main ring beam which holds up the roof) with gables at each end and Trada trusses at approximately 1.60 m centres. (Trada stands for Timber Research and Development Association – these trusses were designed and developed by Trada to enable timber to be used economically. Now most builders prefer to use the Fink type trussed rafters.) The roof was covered in tiles. Beneath the tiles there was a sarking felt layer. (Sarking felt is a secondary layer normally incorporated into new properties to prevent water penetration if a tile is dislodged.)

The roof seemed in good condition. The ridges and verges looked as if they had recently been repointed. The present owner confirmed that this was the case and that the work had been carried out by a competent roofer the previous year. We were led to believe that the reason for having this work carried out was that one ridge tile came off in high winds. The present owner decided to ensure that the remainder were safely bedded.

We gained access to the loft through a trapdoor in the right-hand bedroom (viewed from front). The trap door was well sized being some 700 mm × 600 mm.

The loft was clean but cluttered. The present owner is a smoker and small piles of cigarette butts are apparent in several places. These should be cleaned out. The loft floor was insulated with approximately 2 in (50 mm) of fibreglass. The normal accepted standard for roof insulation is now 4 in (100 mm) but there are recommendations being mooted by the powers that be that 6 in (150 mm) of glassfibre be adopted as the new standard. However, it would be unwise to increase the insulation thickness above 4 in in this case, as the ceiling joists are not deep enough to accommodate any greater thickness. (The glassfibre should not be thicker than the joists otherwise it might obscure the joists from view and make movement in the loft dangerous.)

The main timber sizes were as follows:

Type	Size (mm)	Spacing (mm)	Dist./supports (m)
Rafters	70 × 50	450	1.90 (approx)
Ceiling joists	50 × 100	450	1.70 (approx)
Purlins	50 × 150		1.70 (approx)
Binders	50 × 150		1.70 (approx)

After comparing the sizes with modern tables, we came to the conclusion that by modern standards the timbers were slightly undersized. However, the roof shows no signs of distress and we would therefore conclude that the roof construction is sound. Rust was noted on the bolts and washers to the Trada trusses but we considered the rusting so slight as to be of little importance at this stage.

The roof timbers were tested with a moisture meter and found dry. Judging by the slight dampness noted in the render to the chimney, we believe that the flashing and soakers to the roof may need attention. (Soakers and flashings are purpose-made pieces of lead inserted to fill in a junction between the chimney and the roof.)

While we were carrying out the survey the present owner removed an old water tank from the loft. (He had replaced the tank with a new PVC tank some weeks before and did not wish to leave the old one behind.) We would confirm that the new water tanks seems in good condition. It requires insulating though, and lids to be provided. (No doubt if you mention this to the present owner prior to the sale being completed he might very well carry out the work himself.)

Chimney (1 no.)

When inspected from ground level it appeared that the joints required repointing. It would also be advisable to re-flaunch the top of the chimney. (Flaunching is the cement and sand topping to the chimney.) The lead flashings also require attention. A brick is missing in the loft space and should be replaced.

From observation, it would seem that the chimney is lined with Copex flue liner, which is as it should be.

Main walls of house

No examination has been made of the foundations because this would require digging. The walls were checked for verticality and all was found in order.

There was evidence of a damp proof course (DPC) in the external walls. Because the present owners have laid new flags, the DPC in the walls, in places was slightly less than 150 mm (6 in) above ground level, which is the recommended minimum. However, we could detect no evidence of excessive dampness internally. Below DPC level the bricks were covered in moss indicating that the bricks there are damp. The fact that the moss does not progress above the DPC would also testify to its effectiveness. Approximately ten bricks below DPC level were spalled (i.e. faces were crumbling). These bricks need cutting out and replacing.

We would anticipate that a small jobbing builder could charge in the region of £—— per brick to replace same.

At the rear, the ivy is growing rampantly and we would advise that where it has attached itself to the house, it be cut back otherwise the ivy could force itself between the tiles and into any soft mortar joints. If this were to happen, dampness could enter the property. The ivy will also become a breeding ground for insects and birds.

From discussions with the present owner, we gather that the external walls were foam filled by . . . . . shortly after the house was built. It would be worth asking the present owners for all the documentation on this filling as a guarantee might be transferable to yourself if you buy the property.

NB At one time filling wall cavities was discouraged but nowadays is common. There have been cases where badly installed cavity fill has led to damp penetration and, in some cases, the release of toxic gases.

We have no reasons to believe that there is any problem with the foam filling and it was, according to the present owners, carried out by a reputable firm. We could not find any obvious signs of damp penetration. However, we cannot offer any guarantees on this matter as only laboratory testing, which is beyond the scope of this report, would settle the matter.

Should you wish us to investigate this matter further, we will do so. Without further tests, this is a risk that you will have to accept.

Drainage

We were only shown one manhole by the present owners. We lifted the cover and water was run from the WC and sink. The water flowed freely. The gulleys generally were covered to prevent leaves entering the system.

(Note comments later under discussions with Building Control.)

Windows, doors and timberwork

All the windows, external doors, frames and fascia boards were in excellent condition and only minor areas have signs of slight wood rot.

Gutters and downpipes

The gutters and downpipes were PVC. Other than a light covering of sand in the inverts the gutters were clear.

INTERIOR GENERALLY

Heating installation

The bungalow has gas fired control heating and there were radiators in all major rooms, except the kitchen. The lounge radiator is a continuous skirting type. The boiler is hidden inside a kitchen cabinet and is not well sited. There is also a smell of gas in the cupboard. We would advise that the boiler be inspected by a qualified plumber and resited in a more sensible position. It should also be cleaned. No tests have been carried out on this installation.

Plumbing

The sanitary fittings were tested visually by turning on taps and flushing toilets. All seemed in order. Externally the overflow pipe from the WC has been cut off. This should be replaced so that in the event of overflowing occurring, water does not discharge down the wall.

Electrical installation

The electrical system seemed to be PVC covered and in good condition but no test has been made. With the exception of the kitchen all rooms were well served with electric points and light fittings.

The schedule of points noted was as follows:

	Sockets	Light points
Lounge	4 single	4
Bed. 1	3 single	1
Bed. 2	3 single	1
Bath room	—	1
Hall	2 single	1
Kitchen	2 single	Concealed lighting
	1 double	above ceiling

Internal walls

All internal walls appear to be non-load-bearing partitions/stud partitions. (A stud partition is a framework of timbering covered with plasterboard both sides.) The only disadvantage that we could see to this construction is that in the event of one wishing to hang further cupboards on walls, either a firm bearing on the timber has to be found or cavity toggles used on lighter objects.

Floors

All floors were covered by fitted carpet or stuck finishes. It was therefore only possible to inspect in isolated areas.

The heating pipes are generally run around the edges of the external walls. In order to facilitate this, parts of the existing screed have been hacked up, the pipes installed and the screed replaced. Although burying pipes in floors is not now recommended, it was common practice a few years ago. The reason that burying pipes is not recommended is because in the event of a leak, the screed has to be broken up for access. We had no reason to believe that there were any defects with the system but as hidden pipework cannot be inspected without causing extensive damage, you will have to accept this risk if you buy the property.

Decoration

The property generally was in a reasonable state of decoration.

Bathroom

The fittings were tested and all seemed in good order. See elsewhere for general comments.

Airing cupboard

The airing cupboard contained a lagged cylinder. See elsewhere for general comments.

Hall

See elsewhere for general comments.

Cloaks

See elsewhere for general comments.

Bedrooms

See elsewhere for general comments.

Lounge

See elsewhere for general comments.

Kitchen

The kitchen was fully tiled and had an illuminated ceiling. (This was a series of strip lights hidden by translucent sheeting on a metal grid.) We gather from conversations with the present owner and inspection of receipts that the translucent material is not inflammable and is recommended for kitchen locations by the installer. The kitchen fittings were from a standard range but were in reasonable condition. The gas boiler was hidden inside a cupboard and had no outer casing. We would advise this to be checked by a plumber. (Gas could be smelt.) We were led to believe that except for the washing machine all kitchen fittings are being left.

EXTERNALLY

Rear garden

The garage is detached from the rest of the house, and comprised half brick walls and a flat roof. The roof was covered with a great deal of moss which would indicate that it drains slowly. On top of the roof is a quantity of rotten boards. In our opinion, these should be burnt to prevent possible spread of woodrot. The garage door was tested and seemed in good condition. Judging by the age of the property we would envisage that the felt on the garage roof might need renewal in the near future. The rear garden is well planted with shrubs and trees. Two trees (tree of heaven) were growing by the fence. It is possible if these trees grow too big that they might need cutting down. There is a short alley some 6 ft (2 m) wide at the bottom of the garden which connects to a neighbouring road. We inspected the alleyway and noted that it had two lamp-posts. We would therefore conclude that it is well lit at night and would not encourage undesirables to congregate.

Front garden

The front garden appeared to be open plan. It seemed well planted.
    The flagging and drive is in good condition.

RATING AND PLANNING

We have not made any enquiries regarding rating. We have assumed that your solicitor will be making these enquiries.

BUILDING CONTROL AND COAL BOARD

Discussions with Building Control

In order to carry out as many checks on 6 Snow Road as is reasonbly possible we have spoken to the Building Control Department at Often Council. Local information was given to us about the area on the strict understanding that it was for guidance only, as the ground conditions in the Snow Road area are very variable. After saying that, local officers get to know their areas very well. We were advised that the property probably had a separate drainage system (i.e. separate foul and rainwater drainage). (*Note* When inspecting the drainage, we were only able to find one manhole which seemed to be part of the foul system; we therefore cannot comment on the surface water.)

    Apparently, when some of the houses were built in this area the drains were difficult to install because of lack of 'falls'. Some houses were deliberately built high and the gardens filled with extra soil, in order that normal drains could be installed. As the rear garden has a retaining wall, it would seem likely that this was one of those houses.

    However, according to Building Control, the only real effect that this will have on the property is that if an extension were built, the foundations would need constructing to a depth of about 5 ft.

    We were advised that the Balls Lane area was known for bad ground but that as far as they were aware Snow Road had not had ground problems.

    We were advised however of possible problems with mining. From what we were told, Back Earth Colliery reaches as far as Porcupine Wood. We have requested a mining report from the Coal Board and will forward same to you as soon as it is to hand.

CONCLUSIONS

Subject to obtaining a good report from the Coal Board, and the boiler being resited, the house seems in good condition.

We were particularly asked to inspect the roof to see if it would be possible to carry out a loft conversion.

In our opinion, the ridge was far too low to accommodate the alterations and we do not believe that the Planning Authorities would allow a major alteration to the ridge line of this dwelling.

A. R. WILLIAMS MCIOB

For and on behalf of
Building Services (Technical)

Index